Modern Problems in Nuclear and Neutrino Astrophysics

Arak M. Mathai · Hans J. Haubold

Modern Problems in Nuclear and Neutrino Astrophysics

Second Edition

Arak M. Mathai
Department of Mathematics and Statistics
McGill University
Montreal, ON, Canada

Hans J. Haubold
Vienna International Centre
UN Office for Outer Space Affairs
Vienna, Austria

ISBN 978-3-031-83386-1   ISBN 978-3-031-83387-8 (eBook)
https://doi.org/10.1007/978-3-031-83387-8

This work was supported by Hans Joachim Haubold.

Previously published by Akademie-Verlag Berlin
2nd edition: © The Editor(s) (if applicable) and The Author(s) 2026. This book is an open access publication.

Open Access This book is licensed under the terms of the Creative Commons Attribution 4.0 International License (http://creativecommons.org/licenses/by/4.0/), which permits use, sharing, adaptation, distribution and reproduction in any medium or format, as long as you give appropriate credit to the original author(s) and the source, provide a link to the Creative Commons license and indicate if changes were made.
The images or other third party material in this book are included in the book's Creative Commons license, unless indicated otherwise in a credit line to the material. If material is not included in the book's Creative Commons license and your intended use is not permitted by statutory regulation or exceeds the permitted use, you will need to obtain permission directly from the copyright holder.
The use of general descriptive names, registered names, trademarks, service marks, etc. in this publication does not imply, even in the absence of a specific statement, that such names are exempt from the relevant protective laws and regulations and therefore free for general use.
The publisher, the authors and the editors are safe to assume that the advice and information in this book are believed to be true and accurate at the date of publication. Neither the publisher nor the authors or the editors give a warranty, expressed or implied, with respect to the material contained herein or for any errors or omissions that may have been made. The publisher remains neutral with regard to jurisdictional claims in published maps and institutional affiliations.

This Springer imprint is published by the registered company Springer Nature Switzerland AG
The registered company address is: Gewerbestrasse 11, 6330 Cham, Switzerland

If disposing of this product, please recycle the paper.

A. M. Mathai

This book is published to celebrate the occasion of Prof. Dr. A. M. Mathai's 90th birthday on 28 April 2025

Contents

1 **The Theory of Nuclear Energy Generation of Solar Type Stars** 1
   1.1   Classical Thermodynamics: Polytropic Gas Spheres
         in Convective Equilibrium 1
   1.2   Quantum Theory of Atomic Structure: Opacity 4
   1.3   Quantum Theory of Nuclear Structure: Nuclear Energy
         Generation Rates 5
   1.4   Stellar Models with Nuclear Burning: Nuclear Reaction
         Rates 6
   1.5   The Solar Neutrino Experiment: Solar Neutrino Emission
         Rates 7
   References 10

2 **Mathematical Approaches to Nuclear Astrophysics** 15
   2.1   Introduction: Cosmic Nucleosynthesis of the Elements 15
   2.2   The Thermonuclear Reaction Rate 19
   2.3   Velocity Distribution Function and Nuclear Cross Section:
         Maxwell-Boltzmann Distribution Function 21
   2.4   Nonresonant Neutron Capture Cross Section 22
   2.5   Nonresonant Charged Particle Cross Section 23
   2.6   Resonant Cross Section for Neutrons and Charged Particles 24
   2.7   Parameterizations of Thermonuclear Reaction Rates 25
   2.8   Nonresonant Reaction Rates 28
   2.9   Resonant Reaction Rates 32
   2.10 Series Representations for the Thermonuclear Functions:
         $G_{0,3}^{3,0}(\cdot)$ 37
   2.11 Series Representations for the Thermonuclear Functions:
         $G_{1,3}^{3,0}(\cdot)$ 46
   References 49

3 The Solar Neutrino Problem ... 51
- 3.1 Introduction: The Solar Neutrino Problem ... 51
- 3.2 The Proton-Proton Chain ... 52
- 3.3 An Analytic Model for the Central Region of the Sun ... 56
- 3.4 Solar Thermonuclear Energy Generation: Energy Conservation and Solar Luminosity ... 59
- 3.5 The Thermonuclear Reaction Rate ... 61
- 3.6 The Neutrino Emission Rate ... 63
- 3.7 The Integral for the Solar Nuclear Energy Generation and the Solar Neutrino Fluxes: The General Case of the Basic Integral ... 64
- 3.8 The Solar Case of the Basic Integral ... 66
- 3.9 Fox's H-Function ... 68
- 3.10 Meijer's G-Function ... 69
- 3.11 Analytic Results Connecting Solar Structure Parameters and Solar Neutrino Emission Rates ... 69
- References ... 72

4 Solar Nuclear and Neutrino Astrophysics Research, a Time Line ... 75

5 Nuclear Astrophysics, 2025 Update ... 93
- 5.1 Explicit Evaluation of the Thermonuclear Reaction-Rate Probability Integrals ... 93
- 5.2 Generalization of the Reaction-Rate Probability Integral ... 97
- 5.3 An Extension Through Mathai's Pathway Idea ... 99
- 5.4 The Pathway Extended Reaction-Rate Probability Integral ... 103
- 5.5 Generalized Reaction-Rate Probability Integral in the Real Multivariate Case ... 107
- 5.6 Real Matrix-Variate Case ... 109
- 5.7 Most General Real Matrix-Variate Case ... 110
- 5.8 Generalized Reaction-Rate Integrals in the Complex Multivariate Case ... 113
- 5.9 Complex Matrix-Variate Case ... 114
- 5.10 Most General Form in the Complex Matrix-Variate Case ... 114
- 5.11 Reaction-Rate Probability Integral Through Optimization of Mathai Entropy ... 116
- 5.12 Multivariate Densities in the Real and Complex Domains ... 118
- 5.13 Matrix-Variate Densities in the Real and Complex Domains ... 118
- 5.14 Mellin Convolutions Involving Other Functions ... 119
- 5.15 Generalization to Real Matrix-Variate Case ... 120
- 5.16 Generalization to the Complex Matrix-Variate Case ... 121
- 5.17 Mellin Convolution of a Ratio ... 122
- 5.18 M-Convolution of a Ratio in the Real Matrix-Variate Case ... 123
- 5.19 Fractional Integral of the First Kind in the Complex Matrix-Variate Case ... 124
- References ... 125

6 Neutrino Astrophysics, 2025 Update: The Entropic Approach to Solar Neutrinos 127
6.1 Solar Neutrinos: SuperKamiokande Data 127
6.2 Diffusion Entropy and Standard Deviation: Analysis 127
6.3 Probability Density Function and Differential Equation: Lévy Flights 132
6.4 Discussion 135
References 135

7 Neutrino Astrophysics, 2025 Update: Neutrino Masses and Mixings 137
7.1 Introduction 137
7.2 Modified Dirac and Majorana Neutrino Matrices and Their Distributions 138
7.3 Derivation of the Density of U 141
7.4 Densities in Terms of the Eigenvalues 143
7.5 Exact Marginal Function of the Largest Eigenvalue λ_1 in (7.19) 149
7.6 Exact Marginal Function of λ_p, the Smallest Eigenvalue 150
References 150

Author Index 153

Subject Index 157

Chapter 1
The Theory of Nuclear Energy Generation of Solar Type Stars

1.1 Classical Thermodynamics: Polytropic Gas Spheres in Convective Equilibrium

Shortly after the discovery of the law of conservation of energy by Mayer (1942) and Helmholtz (1847) it was J. R. Mayer who raised the question for the origin of the radiative energy emitted by the Sun. J.R. Mayer's law of conservation of energy (first law of thermodynamics) took into account the energy due to heat. If dQ is the amount of heat energy which is absorbed by the system under question from its surroundings, then this law of conservation of energy is,

$$dQ = dU + dW, \tag{1.1}$$

where dU is the change in internal energy of the system when going from one equilibrium state to another, and dW is the amount of work done by the system on its surroundings. In 1848 J. R. Mayer stated that the source of the solar radiation energy should be the kinetic energy of infalling meteorites. At last a refinement of J. R. Mayer's idea we meet in the hypothesis of Helmholtz (1847) and Lord Kelvin (1861) who were able to show that the gravitational contraction of the Sun itself could be the significant source of radiated energy. However, already on the basis of Lane's (1870) considerations of a stellar configuration in convective equilibrium taking into account internal gravitation and Ritter's (1878) results on the uniform expansion and contraction of gaseous configurations it has been shown that gravitational contraction cannot keep the Sun shining (HELMHOLTZ-KELVIN contraction time scale). This very important conclusion for the theory of internal structure of the Sun is closely related to the macro-structure of a gaseous sphere in stationary equilibrium described by the virial theorem of Clausius (1870) which is also called POINCARE's theorem:

$$2T + \Omega = 0 \tag{1.2}$$

where T is the total kinetic energy of particles, and Ω is the total gravitational potential energy of the system. For an adiabatic process in a gaseous sphere in gravitational equilibrium it follows from (1.1) that the change dU in the internal energy of the sphere is given by

$$dU = -PdV, \quad (1.3)$$

where P is the gas pressure and dV is the change in the volume. Considering a perfect gas with $PV = RT$ one has,

$$\frac{dT}{T} = -(\gamma - 1)\frac{dV}{V}, \quad (1.4)$$

where R is the gas constant and γ is the adiabatic index. Using (1.3) and (1.4) it holds for the perfect gas,

$$U = \frac{2T}{3(\gamma - 1)}, \quad (1.5)$$

which can be written in the light of POINCARE's theorem (1.2),

$$U = -\frac{\Omega}{3(\gamma - 1)}. \quad (1.6)$$

The result in (1.6) goes back to Ritter (1880a). The total energy E is then given by

$$E = U + \Omega = \frac{3\gamma - 4}{3(\gamma - 1)}\Omega. \quad (1.7)$$

From (1.7) it is evident that the gaseous sphere will be unstable against adiabatic pulsations for $\gamma < 4/3$. In his remarkable paper Ritter (1880a) also obtained the result that the period of oscillation of the gaseous sphere is inversely proportional to the square root of its mean density.

As a remark we should mention here that Maxwell (1860) in 1860 obtained the probability distribution of gas particles with velocities between v and $v + dv$,

$$f(v)dv = \left(\frac{2}{\pi}\right)^{\frac{1}{2}} \left(\frac{m}{kT}\right)^{\frac{3}{2}} v^2 \exp\left\{-\frac{mv^2}{2kT}\right\} dv, \quad (1.8)$$

where m is the mass of the particle under question. Already in Ritter's (1879) paper consequences of (1.8) played an important role for the development of convective equilibrium in a gaseous sphere under the influence of its own gravitation. Equation (1.8) follows also directly from the MAXWELL-BOLTZMANN solution L. Boltzmann's (1872) equation found by him in 1872.

For the first time the fundamental differential equation governing the structure of gaseous spheres was given by Ritter (1880) and Lane (1870),

1.1 Classical Thermodynamics: Polytropic Gas Spheres in Convective Equilibrium

$$\frac{1}{r^2}\frac{d}{dr}\left(\frac{r^2}{\rho}\frac{dP}{dr}\right) = -4\pi G\rho, \tag{1.9}$$

which can be derived by consideration of two fundamental equations of gravitational equilibrium,

$$\frac{dM(r)}{dr} = 4\pi r^2 \rho, \quad \text{conservation of mass,} \tag{1.10}$$

$$\frac{dP}{dr} = -\frac{GM(r)\rho}{r^2}, \quad \text{hydrostatic equilibrium.} \tag{1.11}$$

A. Ritter studied the properties of gaseous spheres in convective equilibrium, where pressure P and matter density ρ obey the law

$$P = K\rho^{1+\frac{1}{n}}, \tag{1.12}$$

which are called polytropes of index n. Taking into account (1.12) and (1.9), we are led to the well-known LANE-EMDEN equation,

$$\frac{1}{\xi^2}\frac{d}{d\xi}\left(\xi^2\frac{d\theta}{d\xi}\right) + \theta^n = 0, \tag{1.13}$$

where the solution satisfies the initial condition $\theta(0) = 1$ and $\theta'(0) = 0$. The LANE-EMDEN equation (1.13) can be obtained from (1.9) and (1.12) by substituting $\rho = \lambda\theta^n$, $r = \alpha\xi$, and $\alpha = \left[\frac{(n+1)K}{4\pi G}\lambda^{\frac{1}{n}-1}\right]^{1/2}$. Special cases of (1.13), namely, when $n = 0, 1, 5$, have explicit analytic solutions. The mathematical foundation for the study of the LANE-EMDEN equation (1.13) and of more general equations was made by Fowler (1914a, b, 1930, 1931) in a series of four papers during 1914–1931. However, the theory of polytropic stars governed by the second order nonlinear differential equation (1.13) culminated in Emden's (1907) book. The fundamental point of view adopted at that time was that the energy transport in the interior of a star would be by convective motion. But, already in 1894 Sampson (1894) introduced the concept of energy transfer by radiative rather than by convective processes. This concept had to wait upon further progress in thermodynamics. In 1905 Schuster (1905), E. Rutherford's predecessor in Manchester, applied the idea of radiative energy transfer of an atmosphere which also led to Schwarzschild's (1906) famous paper on the physical state of the Sun's atmosphere. It contains the concept of local thermodynamic equilibrium.

It is generally accepted in the literature that the publication of Emden's (1907) book containing the complete theory of polytropic stars marks the end of the first epoch in the study of the internal constitution of stellar configurations. In the following decades mainly the importance of radiative energy transfer in the interior of stars was realized and the theory of radiative transfer was developed in some detail.

1.2 Quantum Theory of Atomic Structure: Opacity

Although E. Hertzsprung (1905) in 1905 had already recognized the distinction between giant and dwarf stars the full discovery of the HERTZSPRUNG-RUSSELL diagram had to wait until 1913. In 1911–1914 Hertzsprung (1911, 1912) and Russell (1914) were convinced as to the significance of their diagram for the study of the evolution of stars. Indeed the theory of the evolution of stars based on observations first became possible in connection with the study of the internal structure of the stars made by LANE, RITTER, and EMDEN as discussed above.

In Bohr (1913) laid the foundations of the quantum theory of atomic structure by establishing a link between the structure of the atom and PLANCK's quantum of action as given in his trilogy on the constitution of atoms and molecules. It is quite interesting that at about this time the idea was born that the up till now considered mechanical or radioactive energy source did not nearly suffice for supplying the radiation of the Sun. Taking into account that the luminosity of the Sun has not changed significantly since the formation of the Earth some 10^9 years ago, Perrin (1920) and Eddington (1920) in 1920 came to conjecture that in the interior of the Sun subatomic energy must be generated by the conversion of hydrogen into helium. As we shall see in the following the development of the theory of stellar nuclear energy generation is closely connected with the foundations of the theory of astrophysical plasmas. In 1923, for instance, Debye (1923) and E. HÜCKEL had shown that the electrons of a plasma move in such a way as to screen out the COULOMB field of a test charge for distances greater than the DEBYE-HÜCKEL length.

The combination of the theory of radiative equilibrium and BOHR's theory of atomic structure with LANE's, RITTER's, and EMDEN's results was realized by Eddingtion (1916a, b, 1918) (1916–1918) and came up with a more refined theory of the internal structure of the stars. He was the first to apply the concept of radiative equilibrium to the interior of a star and made the assumption that heat is transferred inside a star by radiation whose flow controls the internal temperature distribution. The equations of equilibrium for a star in radiative equilibrium now consist of (1.11) taking into account the gas pressure $P_g = kN_A\rho T/\mu$, and the radiation pressure $p_r = aT^4/3$, the equation of radiative energy transport,

$$\frac{d}{dr}\left(\frac{1}{3}aT^4\right) = -\frac{\chi\rho}{c}\frac{L(r)}{4\pi r^2}, \tag{1.14}$$

and the equation of conservation of energy,

$$\frac{dL(r)}{dr} = 4\pi r^2\rho\epsilon, \tag{1.15}$$

where χ is the measure of the ability of a gas to absorb radiation (opacity), and ϵ is the total amount of heat energy liberated per unit mass in unit time (energy generation rate). While the opacity χ could be fixed by the physical theory of radiative transfer, the quantity ϵ as the nuclear energy generation rate still remained uncertain at that time. The situation was surveyed at the end of the third decade of the century by Eddington's (1926) book During the following decade much work was done on the derivation of detailed stellar models.

1.3 Quantum Theory of Nuclear Structure: Nuclear Energy Generation Rates

The decades after the publication of Eddington's (1926) book are characterized by the breakthrough of quantum mechanics into the physics of stellar interior. In Fowler (1926) made the fundamental discovery that the electron assembly in the white dwarfs must be degenerate in the sense of the FERMI-DIRAC statistics in the same way as shortly thereafter W. Pauli and A. Sommefeld showed to be the case for electrons in metals. He derived the equation of state for degenerate matter, $P = k\rho^{5/3}$, where k is a constant. About this time Wentzel (1926), Kramers (1926), and Brillouin (1926) studied approximate solutions of the SCHRÖDINGER equation for a charged particle in the COULOMB field which later on were named as COULOMB wave functions. Charged particle interactions at low energies as expected in stellar interiors are dominated by the COULOMB barrier penetration factor first discussed in detail by Gamow (1928), Gurney and Condon (1928, 1929). Thus, the theoretical foundations for the stellar nuclear energy generation had been created. In Atkinson (1929) and F.G. Houtermans considered the transmutation of elements arising from proton captures by the help of simple physical considerations. Fowler (1929) and Wilson (1929) developed the resonant penetration of charged particles leading to the distinction of nonresonant and resonant particle interactions in nuclear reactions. It is an important fact that Eddington (1926) in his book succeeded in deriving a relation between the mass M, luminosity L, and the opacity χ from the considerations of a steady state of a star not knowing the dependence of the energy generation rate ϵ on density ρ and temperature T. However, contrary to EDDINGTON's point of view Milne (1930) suggested that the mass, luminosity, and opacity of a stellar model must be taken as independent parameters in the consideration of its steady state, and that the observed mass-luminosity relation should depend upon the intrinsic physics of the energy generating processes and not from deductions of steady state considerations only. During the years 1926–1939 important integral theorems were established by the analysis of the differential equations of the internal structure of stars (1.10), (1.11), and the total pressure as a sum of the gas kinetic pressure p_g and radiative pressure p_r. Those integral theorems permitted the estimation of mean values and values at the centre of the star for all relevant physical variables inside the star (density, pressure, temperature). May be the two most important theorems are due to Vogt (1926) and Russell (1927) as well as to Strömgren (1937). The VOGT-RUSSELL theorem is valid for gaseous stars in radiative equilibrium as well as in convective equilibrium and states that the four first-order ordinary differential equations (1.10), (1.11), (1.14), (1.15), under general assumptions for the three constitutive relations P, χ, ϵ, form a self-sufficient system for the problem of stellar structure if one takes into account, additionally, the boundary conditions: $M(r) = 0$ at $r = 0$; $M(r) = M$, $L(r) = L$, and $P(r) = 0$ at $r = R$. The theorem in question reads that the structure of a star is uniquely determined by its mass and chemical composition, if the pressure P, the opacity χ and the rate of energy generation ϵ are functions of the local values of density ρ, temperature T, and chemical composition X_i only. The main observational consequence of the VOGT-RUSSELL theorem was that different stars in the distribution of stars in the HERTZSPRUNG-RUSSELL diagram have different chemical composition. The theorem of STRÖMGREN states that for a star in radiative equilibrium with negligible radiation pressure the mass-luminosity-radius relation has the form

$$L = \text{constant.} \frac{1}{\chi_0} \frac{M^{5+s}}{R^s} \mu^{7+s}, \qquad (1.16)$$

if the rate of energy generation ϵ and the opacity χ obey the general power laws

$$\epsilon = \epsilon_0 \rho^\alpha T^\nu, \qquad (1.17)$$

$$\chi = \chi_0 \rho T^{-3-s}, \qquad (1.18)$$

where α, ν, s, ϵ_0, and χ_0 are arbitrary constants. The constant in Eq. (1.16) depends only on the exponents α, ν, and s, respectively. Again, the value of STRÖMGREN's theorem (1.16) is to have the dependence of the luminosity L from M and R which allows a comparison of theory and observation. The special feature of STRÖMGREN's theorem is the more realistic dependence of L on the introduced physical assumptions about the rate of energy generation via α and ν and the opacity via s.

Besides the theory of the internal structure of ordinary stars (main sequence stars) Chandrasekhar (1931) realized in 1931 that with increasing relativistic degeneracy the radius of a white dwarf star tends to the limiting value zero at a finite limiting mass of $1.4 M_\odot$ with a slight dependence on chemical composition. The main result was that white dwarf stars are stable or can exist, only for masses $M \leq 1.4 M_\odot$ (chandrasekhar limiting mass). Then, the exact equation of state for a completely degenerate gas has been derived by Chandrasekhar (1935). Much of the material developed for physics of degenerate matter and white dwarf stars as well as the construction of stellar models by analytic methods was summarized in the monograph of Chandrasekhar (1939).

1.4 Stellar Models with Nuclear Burning: Nuclear Reaction Rates

The decades after the publication of Chandrasekhar's (1939) monograph saw the working out of further details of the nuclear reactions and the determination of reaction rates on the basis of laboratory measurements. As an example for the elaboration of the nuclear reaction theory we refer to the analytic treatment of nuclear cross sections as given by Breit (1936) and Wigner's (1936) single resonance formula. The rapid advance of nuclear physics in the thirties of this century enable Weizsäcker (1937) and Bethe (1938) and Critchfield (1938) to work out the nuclear reactions that are possible at temperatures of about 10^6 to 10^8 K in the deep interior of the Sun (CNO cycle). Thus, the problem of stellar energy generation came to a solution. A second possibility for conversion of hydrogen into helium was offered by a reaction chain which Bethe (1939) had studied in 1939 (Proton-Proton chain). Thorough experimental investigations particularly those of Fowler (1984a), of reaction cross sections at low energies have contributed very significantly to our knowledge of nuclear energy generation in stars. One of the first results of the theory of stellar evolution taking into consideration hydrogen burning was that stars remain in the immediate vicinity of the main sequence until a considerable faction ($\approx 10\%$) of the hydrogen is burned. If no mixing takes place between burned and unburned material, the evolutionary track in the effective temperature-luminosity diagram then leads upward and to the right into the region of the red giants, a result obtained by Schönberg (1942) and S. Chandrasekhar as early as 1942. In connection with a more elaborate concept of astrophysical cross section factors given by Salpeter (1952a) (1952) the theory of nuclear burning stars made important advancement. In the deep interior of a star, suppose that a certain part of the hydrogen has been used up then the temperature rises to more than 10^8 K as a result of gravitational contraction. Öpik (1951) and Salpeter (1952b) remarked in 1951/1952 that helium burning sets in to give carbon in accordance with the triple-α-process. The growth of theoretical researches on sellar evolution, starting with the important paper by Hoyle (1955) and Schwarzschild (1955) denoted also a fundamental extension of the LANE-RITTER-EMDEN-EDDINGTON theory of the internal constitution of stars. The assumption hitherto made, that the material inside a star is continually mixed through its evolution, had to be given up in the face of theoretical and observational results. The new concept of the theory of stellar evolution was born, considering the evolution of a star as successive stages of nuclear burning phases. These developments are described in the monograph of Schwarzshild (1958) and Hayashi (1962).

1.5 The Solar Neutrino Experiment: Solar Neutrino Emission Rates

The development of large fast computers at the beginning of the 1960s has had a profound impact on the study of stellar structure. It was no longer necessary to use the techniques of integration by hand discussed in Schwarzschild's (1959) book. The first foundation of the elaborate numerical study of the internal structure and evolution of stars was the method of dividing the structure of a star into many concentric zones and then solving the differential equations of stellar structure in difference form at the boundaries of these zones, introduced by Henyey (1959). The second foundation of this study is Fowler's (FowlerEtAl 1967; HarrisEtAl 1983) nuclear reaction rate systematics as published in a series of 'Handbücher der Kernastrophysik' (1967, 1975, 1983). Acting as a guide FOWLER's nuclear reaction rates are leading the computer from the main sequence stage of a star through the red giant stage up to the final stages of stellar evolution known as white dwarfs and supernovae. As far as the present scope of the theory of internal structure and evolution of stars is concerned a wide range of astrophysical problems emerged which are treated more or less independently from the classical theory of LANE-RITTER-EMDEN-EDDINGTON-CHANDRASEKHAR as described above briefly. To mention some: The theory of black holes and neutron stars, the emission of neutrinos and gravitational waves by stars, pulsating and oscillating stars, close binaries, population III stars, and so on. For an illustrative introduction to the recent history of the theory of the internal constitution and evolution of stars see Kippenhahn (1984) and Iben Jr. (1985).

In the following we are coming back once again to the classical theory of the internal structure of the Sun as established in the thirties. Surely, since the discovery of the nuclear reactions for the solar energy generation by Weizsäcker (1937), Bethe (1938) and Chrichfield (1938), and Bethe (1939) it was known that additionally to the photon a second stable particle with no charge and a rest mass of approximately) zero, that carries away energy in the course of nuclear reaction is emitted by the Sun: the neutrino. Further, two main characteristics of the neutrino were well known, namely, the weak interaction with matter and that neutrinos arise only in the energy generating regions of the stars and carry therefore unlike photons direct evidence of conditions in stellar cores. The first serious attempt to detect neutrinos emitted by the Sun goes back to Davis Jr. (1955) using a radio chemical neutrino detector based on the reaction $^{37}Cl(\nu, e^{-1})\ ^{37}Ar$. This radiochemical method was suggested by Pontecorvo (1946) in 1946. Note that at this time it was generally believed that neutrinos and antineutrinos were equivalent. After a long way of refinements of the $Cl - Ar$-experiment by R. DAVIS Jr. and associates the counting experiments valid for today's experimental and theoretical calculations of the solar neutrino flux started in 1968 by Daavis Jr. (1968), D.S. HARMER, and K.C. HOFFMAN. The net result of the solar neutrino experiment over more than 15 years is a disagreement between theory and observation of about a factor of four. Considering a standard model for the Sun and varying all of the parameters within the limits to be a plausible range of variation it has been shown by Bahcall (1982) that the disagreement still remains if not otherwise a contradiction with observational results appears.

As generally expected the neutrino flux from various nuclear reactions depends strongly on model parameters. But, for a first glance one can predict the number of solar neutrinos reaching the surface of the Earth per cm^2 per s. This total number is approximately,

$$N_\nu = \frac{2L_\odot}{E} \approx 2 \times 10^{38} s^{-1}, \qquad (1.19)$$

where $L_\odot$ is the luminosity of the Sun and E is the energy liberated by the overall net nuclear reaction for the solar nuclear energy generation (Proton-Proton chain),

1.5 The Solar Neutrino Experiment: Solar Neutrino Emission Rates

not including the energy of the two emitted neutrinos. Thus, the solar neutrino flux on Earth is,

$$\phi_\nu = \frac{N_\nu}{4\pi (AU)^2} \approx 6.5 \times 10^{10} \nu \text{ cm}^{-2}\text{s}^{-1}. \tag{1.20}$$

As will be shown in the following Chapters I and II the relation (1.19) can be derived by constructing a simple solar model solving the respective differential equations (1.10), (1.11), (1.15), taking into consideration the boundary conditions as discussed above, and assuming the equation of state for a perfect gas. For that we consider, instead of the phenomenological form of the rate of nuclear energy generation given in (1.17), the relation following from quantum statistical arguments as given by

$$\epsilon_{12} = \frac{1}{\rho} E_{12} r_{12}(\rho, T), \tag{1.21}$$

where r_{12} denotes the thermonuclear reaction rate between nuclear species 1 an 2, E_{12} is the energy liberated in the nuclear reaction under question. The thermonuclear reaction rate r_{12} in (1.21) can be derived from first principles including the distinction between resonant and nonresonant as well as MAXWELL-BOLTZMANNian and non-MAXWELL-BOLTZMANNian reaction rates. At last r_{12} will be a function not only of density ρ, temperature T, and chemical composition X_1, but also of subatomic quantities,

$$r_{12} = f(\rho, T, X_i, Z_1, Z_2, A_1, A_2, S_{12}) \tag{1.22}$$

where Z_i denotes the atomic number, and A_i is the mass number, and S_{12} is the nuclear cross section factor. Then, the nuclear output of the Sun ascribed to the nuclear reaction under question is according to (1.15) and (1.21),

$$L_{12}(R_\odot) = \int_0^{R_\odot} dr \, 4\pi r^2 \rho(r) \epsilon_{12}(r)$$
$$= \int_0^{R_\odot} dr \, 4\pi r^2 E_{12} r_{12}(\rho(r), T(r)). \tag{1.23}$$

The total number of neutrinos due to the nuclear reaction between number species 1 and 2 in the Sun can then be written as

$$N_{12} = \frac{L_{12}(R_\odot)}{E_{12}} = 4\pi \int_0^{R_\odot} dr \, r^2 r_{12}(\rho(r), T(r)). \tag{1.24}$$

The number output of the Sun due to a reaction between nuclear species 1 and 2 as written in (1.22) is a function of stellar model parameters and subatomic quantities:

$$L_{12} = f(R, \rho, T, X_i, Z_1, Z_2, A_1, A_2, S_{12}). \tag{1.25}$$

What concerns the solar neutrino problem is that the relations (1.22) and (1.25) are equivalent to relations (1.17) and (1.16), but, in (1.22) and (1.25) the subatomic physics is taken into account. All cases of thermonuclear reaction rates (1.22) are derived in Chapter I, whereas the evaluation of (1.25) is given in Chap. 2 in detail.

References

Bahcall, J.N., Huebner, W.F., Lubow, S.H., Parker, P.D., Ulrich, R.K.: Standard solar models and the uncertainties in predicted capture rates of solar neutrinos. Rev. Mod. Phys. **54**(767–799), 65 (1982)

Barnes, C.A., Clayton, D.D., Schramm, D.N. (eds.): Essays in Nuclear Astrophysics. Cambridge University Press, Cambridge-London-New York-New Rochelle-Melbourne-Sydney (1982)

Bethe, H.A.: Energy generation in stars. In Les Prix Nobel 1967, Almqvist and Wiksells, Stockholm 1967 (for a German translation cp. also Die Naturwissenschaften **55**, 405–413 (1968)

Bethe, H.A.: Energy production in stars. Phys. Rev. **55**(103), 434–456 (1939)

Bethe, H.A., Critchfield, C.L.: The formation of deuterons by proton combination. Phys. Rev. **54**, 248–254 (1938)

Bohr, N.: On the constitution of atoms and molecules. The London, Edinburgh, and Dublin Philosophical Magazine and Journal of Science, vol. XXVI.-Sixth Series, Art. I. pp. 1–25, Art. XXXVII, pp. 476–502, Art. LXXIII., pp. 857–875 (1913)

Boltzmann, L.: Weitere Studien über das Wärmegleichgewicht unter Gasmolekülen. Sitzungsberichte der Kaiserlichen Akademie der Wissenschaften Wien, Mathematisch-Naturwissenschaftliche Classe, zweite Abtheilung **66**, 275–370 (1872)

Breit, G., Wigner, E.: Capture of slow neurons. Phys. Rev. **49**, 519–531 (1936)

Brillouin, M.L.: Remarques sur la mécanique ondulatoire. J. de: hysique et la Radium **7**, 353–368 (1926)

Candrasekhar, S.: An Introduction to the Study of Stellar Structure. The University of Chicago Press, Chicago (1939), and Dover Publications, Inc., New York (1967)

Chandrasekhar, S.: The highly collapsed configurations of a stellar mass. Mon. Not. R. Astron. Soc. (Lond.) **91**, 456–466 (1931)

Chandrasekhar, S.: The highly collapsed configurations of a stellar mass (second paper). Mon. Not. R. Astron. Soc. (Lond.) **95**, 207–225 (1935)

Chandrasekhar, S.: On stars, their evolution and their stability. Rev. Mod. Phys. **56**, 137–147 (1984)

Clausius, R.: Über einen auf die Wärme anwendbaren mechanischen Satz. Verhandlungen des naturhistorischen Vereins der preussischen Rheinlande und Westfalens, Bonn, Dritte Folge: 7. Jahrgang **27**, 114–119 (1870)

Davis, R., Jr.: Attempt to detect the antineutrinos from a nuclear reactor by the $^{37}Cl(\bar{\nu}, e^-)\,^{37}Ar$ reaction. Phys. Rev. **97**, 766–769 (1955)

Davis, R., Jr., Harmer, D.S., Hoffman, K.C.: Search for neutrinos from the sun. Phys. Rev. Lett. **20**, 1205–1209 (1968)

de Broglie, M.L.: Les principles de la nouvelle mécanique ondulatoire. Le J. de Physique et le Radium **7**, 321–337 (1926)

Debye, P., Hückel, E.: Zur Theorie der Elekrolyte: I. Gefrierpunktserniedrigung und verwandte Erscheinungen; II. Das Grundgesetz für die elektrische Leitfähigkeit. Physikalische Zeitschrift **24**, 185–206, 305–325, 428 (1923)

Eddington, A.S.: Further notes on the radiative equilibrium of the stars. Mon. Not. R. Astron. Soc. (Lond.) **77**, 596–612 (1916b)

Eddington, A.S.: On the radiative equilibrium of the stars. Mon. Not. R. Astron. Soc. (Lond.) **77**, 16–35 (1916a)

References

Eddington, A.S.: Reports of the Meetings of the British Association for the Advancement of Science, 45 (1920)

Eddington, A.S.: The internal Constitution of the Stars. Cambridge University Press, Cambridge (1926) (for a German translation see A.S. Eddington: Der innere Aufbau der Sterne, Verlag von Julius Springer, Berlin 1928)

Eddington, A.S.: On the radiative equilibrium of the stars. Mon. Not. R. Astron. Soc. (Lond.) **79**, 22–23 (1918)

Emden, R.: Gaskugeln - Anwendungen der mechanischen Wärmetheorie auf kosmologische und meterorologische Probleme. Verlag von B.G. Teubner, Leipzig und Berlin (1907)

Fowler, W.A., Caughlan, G.R., Zimmerman, B.A.: Thermonuclear reaction rates. I. Annu. Rev. Astron. Astrophys. **5**, 525–570 (1967). II. Annu. Rev. Astron. Astrophys. **13**, 69–112 (1975)

Fowler, R.H., Wilson, A.H.: A detailed Study of the "Radioactive decay" of, and the penetration of α-particles into, a simplified one-dimensional nucleus. Proc. R. Soc., Ser. A **124**, 493–501 (1929)

Fowler, W.A.: Experimental and theoretical nuclear astrophysics: the quest for the origin of the elements. Rev. Mod. Phys. **56**, 149–179 (1984a)

Fowler, W.A.: Experimental and theoretical nuclear astrophysics: the quest for the origin of the elements. Rev. Mod. Phys. **56**, 149–179 (1984b)

Fowler, R.H.: The form near infinity of real, continuous solutions of a certain differential equation of the second order. Q. J. Math. **45**, 289–350 (1914a)

Fowler, R.H.: Some results on the form near infinity of real continuous solutions of a certain type of second order differential equations. Proc. Lond. Math. Soc. **13**, 341–371 (1914b)

Fowler, R.H.: On dense matter. Mon. Not. R. Astron. Soc. (Lond.) **87**, 114–122 (1926)

Fowler, R.H.: The solution of Emden's and similar differential equations. Mon. Not. R. Astron. Soc. (Lond.) **91**, 63–91 (1930)

Fowler, R.H.: Further studies of Emden's and similar differential equations. Q. J. Math. (Oxf.) **2**, 259–288 (1931)

Frank-Kameneckij, D.A.: Physikalische Prozesse im Inneren der Sterne, (im russisch). Fizmatgiz, Moskau (1959)

Gamow, G.: Thirty Years that Shook Physics. Doubleday and Co., Inc., New York (1966), and Dover Publications, Inc., New York (1985)

Gamow, G.: Zur Quantentheorie des Atomkerns. Z. Phys. **51**, 204–212 (1928)

Gurney, R.W., Condon, E.U.: Wave mechanics and radioactive disintegration. Nature (London) **122**, 439 (1928)

Gurney, R.W., Condon, E.U.: Quantum mechanics and radioactive disintegration. Phys. Rev. **33**, 127–140 (1929)

Harris, M.J., Fowler, W.A., Caughlan, G.R., Zimmerman, B.A.: Thermonuclear reaction rates III. Ann. Rev. Astron. Astrophys. **21**, 165–176 (1983)

Haubold, H.J., John, R.W.: Der gegenwärtige Stand der Theorie und der analytischen Auswertung von nichtresonanten thermonuklearen Reaktionsraten. Astron. Nachr. **303**, 161–187 (1982)

Hayashi, C., Hōshi, R., Sugimoto, D.: Evolution of the stars: supplement of the progress of theoretical physics **22**, 1–183 (1962)

Helmholtz, G.: Ueber die Erhaltung der Kraft. Eine physikalische Abhandlung. Erschien zu Berlin bei G. Reimer 1847. Wiederabgedruckt im H. HELMHOLTZ: Wissenschaftliche Abhandlungen. Erster Band. Johann Ambrosius Berth, Leipzig 1882. Nr. II, S. 12–75. The discovery of the universality of the conservation of energy cannot be attributed to only one person; for a discussion of that topic see L. KOENIGSBERGER: Hermann von Helmholtz. Erster Band. Friedrich Vieweg und Sohn, Braunschweig (1902)

Henyey, L.G., Le Levier, R., Lenée, R.D.: Evolution of main-sequence stars. Astrophys. J. **129**, 2–19 (1959)

Hertzsprung, E.: Zur Strahlung der Sonne. Zeitschrift für Wissensechaftliche Photographie, Photophysik und Photochemie **III**, 429–442 (1905)

Hertzsprung, E.: Über die Verwendung photographischer effektiver Wellenlängen zur Bestimmung von Farbenäquivalenten. Publikationen des Astrophysikalischen Observatoriums zu Potsdam **22**(63), 1–36 (1911)

Hoyle, F., Schwarzschild, M.: On the evolution of Type II stars. Astrophys. J. **2**, 1–40 (1955)

Iben, I., Jr.: The life and times of an intermediate mass star - in isolation/in a close binary. Q. J. R. Astron. Soc. **26**, 1–39 (1985)

Kippenhahn, R.: Vom Lebenslauf der Sterne. Nova Acta Leopoldina. Im Auftrage des Präsidiums herausgegeben von J. H. SCHARF. Neue Folge. Nummer 260, Band **57**, 1–32 (1984)

Kramers, H.A.: Wellenmechanik und halbzahlige Quantisierung. Z. Phys. **39**, 828–840 (1926)

Lane, J.H.: On the theoretical temperature of the sun, under the hypothesis of a gaseous mass maintaining its volume by its internal heat, and depending on the laws of gases as known to terrestrial experiments. Am. J. Sci. Arts. Second Series, **L**. (Whole Number, C.). Number CXLVIII. Art. IX. pp. 57–74 (1870)

Lang, K.R.: Astrophysical formulae - a compendium for the physicists and astrophysicist. Second Corrected and Enlarged Edition. Springer, Berlin-Heidelberg-New York (1980)

Lord Kelvin (Thomson, W.).: Physical considerations regarding the possible Age of the Sun's Heat. Reports of the Meeting of the British Association for the Advancement of Science, Notices and Abstracts of Miscellaneous Communications to the Sections, pp. 27–28 (1861)

Maxwell, J.C.: Illustrations of the dynamical theory of gases - Part. I. On the motions and collisions of perfectly elastic spheres. The London, Edinburgh, and Dublin Philosophical Magazine and Journal of Science, vol. XIX. Fourth Series. 1860, Art. V, pp. 19–32

Mayer's, R.J., work see Mittasch, A.: Kraft. Leben. Geist - Eine Lese aus Robert Mayers Schriften. Herausgegeben im Namen der Kaiserlichen Leopoldinisch-Carolinisch Deutschen Akademie der Naturforscher von E. Abderhalden. Halle (Saale) S. 1–51 (1942)

Menzel, D.H., Bhatnagar, P.L., Sen, H.K.: Stellar Interiors. The International Astrophysics Series Volume Six. Chapman and Hall Ltd., London (1963)

Milne, E.A.: The analysis of stellar structure. Mon. Not. R. Astron. Soc. (Lond.) **91**, 4–55 (1930)

Öpik, E.J.: Stellar models with variable composition: II. Sequences of models with energy generation proportional to the fifteenth power of temperature. In: Proceedings of the Royal Irish Academy, Dublin, Section A. - Mathematical, Astronomical, and Physical Science, vol. 54, pp. 49–77 (1951)

Pais, A.: Inward bound of matter and forces in the physical world. Oxford, and Oxford University Press, New York, Clarendon Press (1986)

Perrin, J.: Revue du mais **21**, 113 (1920)

Pontecorvo, B.: Inverse β decay. National Research Council Canada. Report PD-205 of the Chalk River Laboratory, pp. 1–7 (1946). Reprinted in Proceedings of the Solar Neutrino Conference, Reines, F., Trimble, V. (eds.), University of California, Irvine (1972), Appendices

Ritter, A.: Untersuchungen Über die Höhe der Atmosphäre und die Constitution der gasförmigen Weltkörper. Wiedemanns Annalen der Physik und Chemie. Neue Folge **7**, 304–317 (1879)

Ritter, A.: Untersuchungen über die Höhe der Atmosphäre und die Constitution der gsförmigen Weltkörper. Wiedemanns Annalen der Physic und Chemie. Neue Folge **11**, 332–344 (1880)

Ritter, A.: Untersuchungen über die Höhe der Atmosphäre und die Constitution gasförmiger Weltkörper. Wiedemanns Annalen der Physik und Chemie. Neue Folge **5**, 543–558 (1878)

Ritter, A.: Untersuchungen über die Höhe der Atmosphäre und die Constitution gasförmiger Weltkörper. Wiedemanns Annalen der Physik und Chemie. Neue Folge **8**, 157–183 (1880a)

Russell, H.N.: Relation between the spectra and other characteristics of stars. Proc. Am. Philos. Soc. **51**, 569–579 (1912)

Russell, H.N.: Relations between the spectra and other characteristics of the stars. Popul. Astron. XXI(275–294), 331–351 (1914)

Russell, H.N.: Astronomy - A Revision of Young's Manual of Astronomy, pp. 902–932. Part II. Astrophysics and Stellar Astronomy. Ginn and Company, Boston (1927)

Salpeter, E.E.: Nuclear reactions in the stars I. Proton-Proton-Chain. Phys. Rev. **88**, 547–553 (1952a)

Salpeter, E.E.: Nuclear reactions in stars without hydrogen. Astrophys. J. **115**, 326–328 (1952b)

References

Sampson, R.A.: On the rotation and mechanical state of the sun. Mem. R. Astron. Soc. **51**, 123–183 (1894)

Schönberg, M., Chandrasekhar, S.: On the Evolution of mainsequence stars. Astrophys. J. **96**, 161–172 (1942)

Schuster, A.: Radiation through a foggy atmosphere. Astrophys. J. **21**, 1–22 (1905)

Schwarzschild, M.: Structure and evolution of the Stars. Princeton University Press, Princeton (1958), and Dover Publications, Inc., New York (1965)

Schwarzschild, K.: Über das Gleichgewicht der Sonnenatmosphäre. Nachrichten von der Königlichen Gesellschaft der Wissenschaften zu Götingen. Mathematisch-Physikalische Klasse 41–53 (1906)

Strömgren, B.: Die Theorie des Sterninnern und die Entwicklung der Sterne. Ergebnisse der Exakten Naturwissenschaften **16**, 465–534 (1937)

Stuewer, R., (ed.): Nuclear Physics in Retrospect - Proceedings of a Symposium on the 1930s. University of Minnesota Press, Minneapolis (1979)

Vogt, H.: Die Beziehung zwischen den Massen und den absoluten Leuchtkräften der Sterne. Asronomische Nachrichten **226**, 304–310 (1926)

Weizsäcker, C.F.: Über Elementumwandlungen im Inneren der Sterne. I. Physikalische Zeitschrift **38**, 176–191 (1937). II. **39**, 633–646 (1938)

Wentzel, G.: Eine Verallgemeinerung der Quantenbedingungen für die Zwecke der Wellenmechanik. Z. Phys. **38**, 518–529 (1926)

Zur Frage der Aufbaumöglichkeit der Elemente im Sternen: D'e. Atkinson, R., Houtermans, F.G. Z. Phys. **54**, 656–665 (1929)

Open Access This chapter is licensed under the terms of the Creative Commons Attribution 4.0 International License (http://creativecommons.org/licenses/by/4.0/), which permits use, sharing, adaptation, distribution and reproduction in any medium or format, as long as you give appropriate credit to the original author(s) and the source, provide a link to the Creative Commons license and indicate if changes were made.

The images or other third party material in this chapter are included in the chapter's Creative Commons license, unless indicated otherwise in a credit line to the material. If material is not included in the chapter's Creative Commons license and your intended use is not permitted by statutory regulation or exceeds the permitted use, you will need to obtain permission directly from the copyright holder.

Chapter 2
Mathematical Approaches to Nuclear Astrophysics

2.1 Introduction: Cosmic Nucleosynthesis of the Elements

According to the current understanding of the evolution of the Universe the presently observed state of the Universe is the result of expansion from an extremely dense and extremely hot singular origin. Describing that evolution of the Universe by means of the standard cosmological model, based on the 'big-bang' hypothesis, cosmological nucleosynthesis occurs at the appropriate temperature in the course of expansion and goes on until the decreasing temperature stops nuclear reactions. No significant cosmological nucleosynthesis beyond helium-4 occurred due to the instability gaps at mass number 5 and mass number 8 as well as the constraints set by the present universal density and temperature. However, it is well-known that 'big-bang' nucleosynthesis results in the production of lithium-7 in amounts comparable to the solar system abundance of this nucleus. Some general features of the cosmological nucleosynthesis are listed in Table 2.1 and Figs. 2.1 and 2.2.

Stars are born out of interstellar gas and are thought to be the site of stellar nucleosynthesis during their lifetime. It is now believed that most of the heavy elements are cooked in successive generations of stars. As can be seen in Table 2.1 and Figs. 2.1 and 2.2, within 10^9 years after the big-bang, stars are formed and stellar nucleosynthesis activities can take place in these stars. It was the study of the detailed plot of the number distribution of the cosmic elements as a function of atomic weight (cp. Figs. 2.1 and 2.2) and COULOMB barrier penetration considerations which led Burbidge et al. (1957) to postulate the basic nucleosynthetic processes in stars as shown in Table 2.1. Hydrogen burning, where four hydrogen nuclei are converted into one helium-4 nucleus, can occur via two different sets of nuclear reactions: the proton-proton chain and the CNO cycles. The most important feature of the CNO cycles is the conversion of carbon-12 and oxygen-16 into nitrogen-14. Helium burning bridges the mass number 5 and mass number 8 instability gaps via a three-particle reaction: 3 helium-4 $\to$ carbon 12. Helium burning includes the conversion of helium-4 into heavier elements such as carbon-12, oxygen-16, and neon-20. Helium burning is also considerd as the site of the slow neutron capture processes (s-process). Carbon burn-

Table 2.1 Basic facts of cosmic nucleosynthesis (cp. Burbidge et al. (1957), Fowler et al. (1983))

	Stage of nucleo-synthesis	Temperature Density Time scale	Main product of the nucleo-synthetic processes	Site of nucleosynthesis
Cosmological nucleosynthesis starts three minutes after the "big bang"		$T \approx 10^9$ K $\varrho \approx 1$ gcm^{-3}	$\boxed{^1H}$, ^2H, ^3He, ^7Li, $\boxed{^4He}$	Cosmology "big bang"
Stellar nucleosynthesis (hydrostatic) starts 10^9 years after the "big bang" Charged particle and neutron capture reactions	1. H burning	$T \gtrsim 10^7$ K $\varrho \approx 10^2$ gcm^{-3} $t \approx 10^{10}$ years	Proton-proton chain in low mass stars $^1H \rightarrow \boxed{^4He}$ CNO cycles in high mass stars $^1H \rightarrow \boxed{^4He}$ $^{12}C, ^{16}O \rightarrow ^{14}N$	Stellar physics "Main sequence star"
	2. He burning s-process	$T \gtrsim 10^8$ K $\varrho \approx 10^3$ gcm^{-3} $t \approx 10^7$ years	Production of $\boxed{^{12}C, ^{16}O}$ via (α, γ) conversion of ^{14}N to ^{18}O, ^{21}Ne, ^{22}Ne $3\,^4He \rightarrow ^{12}C$ to bridge the instability gaps at masses 5 and 8 production of neutrons by (α, n) reactions on ^{21}Ne, ^{22}Ne	"Red giant star"
	3. C burning	$T \gtrsim 6 \times 10^8$ K $\varrho \approx 10^5$ gcm^{-3} $t \approx 10^5$ years	$^{12}C + ^{12}C \rightarrow \alpha + ^{20}Ne, p + ^{23}Na, n + ^{24}Mg$ $\boxed{16 \leq A \leq 28}$	"Onion skin star" "Pre-supernova stage"
	4. Ne burning transient stage	$T \gtrsim 10^9$ K $\varrho \approx 10^6$ gcm^{-3}	$^{20}Ne(\gamma, \alpha)\,^{16}O$, $^{20}Ne(\alpha, \gamma)\,^{24}Mg$ $^{24}Mg(\alpha, \gamma)\,^{28}Si$	
	5. O burning	$T \gtrsim 1.5 \times 10^9$ K $\varrho \approx 10^7$ gcm^{-3} $t \approx 10^5$ years	$^{16}O + ^{16}O \rightarrow$ Si, S, Ar, Ca, ... $\boxed{16 \leq A \leq 28}$	
	6. Si burning	$T \gtrsim 3...4 \times 10^9$ K $\varrho \approx 10^8$ cm^{-3} $t \approx 1$ s	Si $\rightarrow \alpha, p, n \rightarrow$ iron peak elements $\boxed{28 \leq A \leq 60}$ inverse (α, γ) reactions: $^{28}Si(\gamma, \alpha)\,^{24}Mg(\gamma, \alpha)\,^{20}Ne\,(\gamma, \alpha)\,^{16}O$ $^{16}O(\gamma, \alpha)\,^{12}C(\gamma, 3\alpha)$	

2.1 Introduction: Cosmic Nucleosynthesis of the Elements

Table 2.1 (continued)

	Stage of nucleo-synthesis	Temperature Density Time scale	Main product of the nucleo-synthetic processes	Site of nucleosynthesis
Stellar nucleo-synthesis (explosive) during stellar collapse to a neutron star or black hole	7. s-process	$T \gtrsim 5 \times 10^8$ K $\varrho \approx 10^2$ gcm^{-3} $t \approx 10^3 \ldots 10^7$ years	long time scale; $\varphi_n \approx 10^5$ cm^{-3} (n, γ) reactions on seed nuclei $\boxed{A \geq 60}$	"Supernova"
	8. r-process	$T \gtrsim 10^{10}$ K $\varrho \approx 10^9$ gcm^{-3} $t \approx 10 \ldots 100$ sec	short time scale: $\varphi_n \approx 10^{20}$ cm^{-3} (n, γ) reactions on seed nuclei $\boxed{A \geq 60}$	
	9. p-process	$T \gtrsim 2 \times 10^9$ K $\varrho \approx 10^8$ gcm^{-3} $t \approx 10 \ldots 100$ sec	Production of $\boxed{\text{heavy proton-rich nuclei}}$ by (γ, n) reactions	
Cosmic ray nucleo-synthesis (spallation)	10. l-process		^{6}Li, ^{9}Be, ^{10}B, ^{11}B Production of light elements by spallation of C, N, O by cosmic ray α, p	Cosmic ray physics α

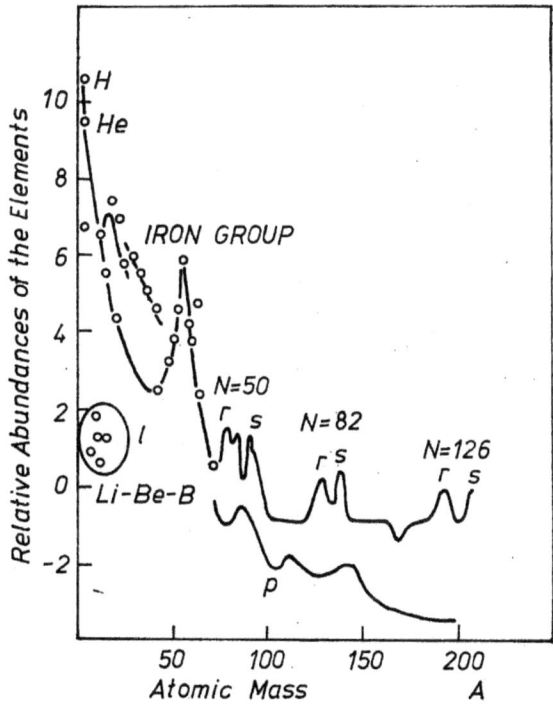

Fig. 2.1 The standard abundance distribution of the elements in the Universe (Si $\equiv 10^6$) (Burbidge et al. (1957))

Fig. 2.2 Abundances of s, r, and *p* process elements

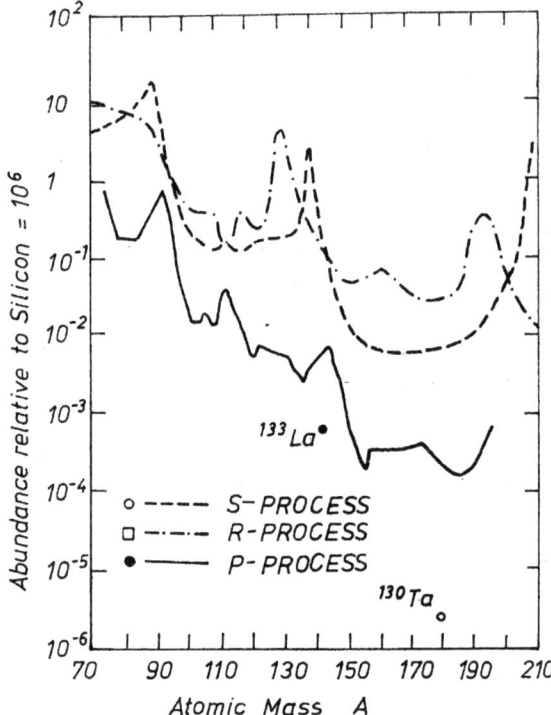

ing consists of the fusion of carbon nuclei themselves to build nucleides in the mass range $16 \leq A \leq 28$. Neon burning is a transient stage to photodisintegrate neon-20 into oxygen-16 and alpha particles. Again, oxygen burning produces elements in the mass range $16 \leq A \leq 28$, but at temperatures still higher than in carbon burning. In silicon burning a strong flux of neutrons, protons, and alphas are liberated mainly from silicon-28. These articles are captured by other nuclei to produce elements up to the iron peak elements. Thus, in these equilibrium processes between synthesis and disintegration, elements with the highest nuclear stability are produced. With the occurrence of silicon burning processes inside a star the so-called stationary stellar nucleosynthesis is finished. The core of the star consisting of iron peak elements will collapse never being stopped by nuclear energy generating reactions. During the collapse a shock wave is generated which propagates out through the star, adiabatically compressing, and thereby heating, the stellar matter through which it passes. The temperature in each nuclear burning shell of the star is raised suddenly and leads to explosive stellar nucleosynthesis in the various shells of the star. The final stages of this collapse and the explosive nucleosynthesis will be observed as supernova.

It is one of the most important features for stellar nucleosynthesis that the binding energy per nucleon decreases with increasing mass number for nucleides beyond the iron peak ($A > 60$), and that these nuclei have large COULOMB barriers (cp. Figs. 2.1 and 2.2). Thus, they are not be formed by charged particle reactions. It is believed that most of these heavy elements are produced by neutron capture reactions which start with the iron group nuclei. These nuclei can be divided into two groups: if the flux of neutrons is weak, most chains of neutron captures will include only a few captures before the beta decay of the product nucleus. Because the neutron capture lifetime is slower than the beta decay lifetime, this kind of neutron capture is called s-process. In the opposite case, if there is a strong neutron flux, the neutron-rich elements will be formed by the rapid neutron capture process, where the neutron capture lifetime is much less than the beta decay lifetime (r-process). There are some very rare nuclei which are located on the proton-rich side of the valley of stability. It is believed that they are produced by the p-process in which photodisintegration occurs on s-process nuclei (Figs. 2.1 and 2.2).

Only to mention it, there are some very light, low abundance nuclei (hydrogen-2, helium-3, beryllium, lithium, bor). They are not produced in sufficient quantities in the cosmological nucleosynthesis and are immediately destroyed by thermonuclear reactions in the course of sellar nucleosynthesis. These nuclei are ascribed to l-processes: spallation reactions on carbon, nitrogen, and oxygen nuclei, by protons or alphas, in the high energy and low energy cosmic rays.

The present generally accepted framework of nuclear astrophysics can already be found in the classical paper by Burbidge et al. (1957) and most recently reviewed in the Nobel lecture of Fowler (1983). The laboratory approach to nuclear astrophysics, what concerns the element synthesizing nuclear reactions studied to a large extent in nuclear laboratories, is presented in the 'Handbücher der Astrophysik' by Fowler et al. (1967, 1975). The outcome of model computations with the best available nuclear input for the cosmological and stationary as well as explosive stellar nucleosynthesis is summarized in articles in 'Nucleosynthesis' edited by Biswas et al. (1980). The aim of the present chapter is to contribute to the problem of the parametrization of nuclear reaction rates as described in Fowler's paper (1983, Table I on p. 154 and Table II on P. 155; ep. also Bethe 1967). In this connection we try to continue the earlier investigations of Atkinson and Houtermans (1929), Gamow and Teller (1938), Salpeter (1952), Bahcall (1966), and Critchfield (1972) as recently considered in Haubold and John (1982) and Haubold and Mathai (1984). The examples for the analytic representation of nuclear reaction rates presented in the following sections should illustrate the field of nuclear astrophysics from the more mathematical point of view. They are worked out in the spirit of the remarks in Fowler's (1974) George Darwin lecture: "... each reaction rate, measured accurately or computed from good systematics, is a diagnostic tool in determining the astrophysical circumstances and sites of nucleosynthesis and energy generation."

2.2 The Thermonuclear Reaction Rate

Before we shall come to the mathematical methods for the analytic representation of nuclear reaction rates it will be useful to discuss some basic nuclear astrophysical relations containing the thermonuclear reaction rate. Most of the formalism for the tabulations of the reaction rate of interacting particles under cosmological or stellar conditions has been presented by Fowler et al. (1967, 1975).

For the most common case, in which two particles in the initial channel (1 and 2) form two particles in the final channel (3 and 4),

$$1 + 2 \rightarrow 3 + 4 + E_{12}, \tag{2.1}$$

the reaction rate r_{12} is defined by (Fowler et al. 1967, 1975; Haubold and Mathai 1984)

$$r_{12} = \left(1 - \frac{1}{2}\delta_{12}\right) n_1 n_2, <\sigma v>_{12}, \ [r_{12}] = cm^{-3}s^{-1}, \tag{2.2}$$

where n_1 and n_2 are the number densities of nuclei 1 and 2, respectively, δ_{12} is the KRONECKER symbol, E_{12} is the energy release ($E_{12} = (m_1 + m_2 - m_3 - m_4)c^2$), c denotes the velocity of light. The KRONECKER delta is introduced to avoid double counting in the reaction if 1 and 2 being identical. In (2.2) the quantity $<\sigma v>$ is the thermally averaged product of the cross section, σ, for the reaction, and relative velocity, v, of the interacting particles, 1 and 2, which is in fact the most important quantity for the analytic representation of the thermonuclear reaction rate (2.2) as we shall see in Sect. 2.3. For a gas of mean density, ρ, the number density, n_i, of particle, i, can be expressed in terms of its mass fraction, X_i, by the relation

$$n_i = \rho N_A \frac{X_i}{A_i}, i = 1, 2, 3, 4, \ [n_i] = cm^{-3}, \tag{2.3}$$

where N_A stands for AVOGADRO's constant, and A_i is the atomic mass of particle i in atomic mass units (Note $\sum_i X_i = 1$).

The mean lifetime, $\tau_2(1)$, of particle 1 for interaction with particle 2 can be given by

$$\lambda_2(1) = \frac{1}{\tau_2(1)} = n_2 <\sigma v>_{12} = \rho N_A \frac{X_2}{A_2} <\sigma v>_{12}$$

$$= -\frac{1}{n_1}\left(\frac{dn_1}{dt}\right)_2 = -\frac{1}{X_1}\left(\frac{dX_1}{dt}\right)_2, \ [\tau_2(1)] = s^{-1}, \tag{2.4}$$

where $\lambda_2(1)$ is the decay rate of 1 for interaction with 2. The definition of the mean lifetime of a nucleus for interaction with a nucleus in (2.4) shows directly the connection of the quantity $<\sigma v>$ with kinetic equations for the production and destruction of the respective nucleus (Haubold and Mathai 1984).

The energy generation rate, ϵ_{12}, for the reaction (2.1) is defined by

$$\epsilon_{12} = \frac{1}{\rho} r_{12} E_{12}, \ [\epsilon_{12}] = \text{erg } g^{-1}s^{-1}, \tag{2.5}$$

where E_{12} is the energy given off in one single reaction (2.1).

2.3 Velocity Distribution Function and Nuclear Cross Section ...

By definition, the quantity $<\sigma v>$ in (2.2) arises from an integral over the respective cross section of the reaction, times relative velocity of the reacting particles, times the distribution function of the relative velocities of the particles:

$$<\sigma v> = \int_0^\infty d^3v \, \sigma(v) \, v f(v) = \int_0^\infty dE \, \sigma(E) \left(\frac{2E}{\mu}\right)^{\frac{1}{2}} f(E), \, [<\sigma v>] = cm^3 \, s^{-1}, \quad (2.6)$$

where $d^3v = 4\pi v^2 dv$; the kinetic energy of the particles in the center-of-mass system is $E = \mu v^2/2$, the reduced mass of the particles is denoted by $\mu = m_1 m_2/(m_1 + m_2)$, the reaction cross section is $\sigma(v)$ and $\sigma(E)$, respectively. Equation (2.6) contains what we expected to be the basic quantities for the description of reactions between nuclear particles going on in the very hot and intermediate dense plasma at some stage of the evolution of the Universe and in the deep interior of stars: for quantummechanical reasons the cross section of the particle reaction and for reasons of statistical mechanics the distribution function of the velocities of the reacting particles.

The time reversal invariance of the strong, electromagnetic, and weak interactions leads to an important relation between the cross sections for the forward and backward nuclear reactions which should be noted here (cp. Blatt and Weisskopf 1959):

$$\frac{\sigma_{34}}{\sigma_{12}} = \frac{(1+\delta_{34}) \, g_1 g_2}{(1+\delta_{12}) \, g_3 g_4} \frac{A_1 A_2 E_{12}}{A_3 A_4 E_{34}}, \quad (2.7)$$

which reads for the quantity $<\sigma v>$ in (2.6):

$$\frac{<\sigma v>_{34}}{<\sigma v>_{12}} = \frac{(1+\delta_{34}) g_1 g_2}{(1+\delta_{12}) g_3 g_4} \left(\frac{A_1 A_2}{A_3 A_4}\right)^{\frac{3}{2}} \exp\left\{-\frac{E_{34}-E_{12}}{kT}\right\}, \quad (2.8)$$

where the g_i's denote spin statistical weights $g_i = (2s_i + 1)$, $i = 1, 2, 3, 4$. At high temperatures many nuclear states are populated and the g_i's become partition functions in the expression (2.8). Equation (2.7) underlines the principle of microscopic reversibility, (2.8) represents the principle of detailed balance (see Haubold and Mathai 1984). All relations given above are thought to be valid for reactions in a non-degenerate gas and for non-relativistic particle velocities.

2.3 Velocity Distribution Function and Nuclear Cross Section: Maxwell-Boltzmann Distribution Function

All the analytic expressions for astrophysically relevant nuclear reaction rates given in the tabulations of Fowler et al. (1967, 1975) underline the hypothesis that the distribution of the relative velocities of the reacting particles always remains Maxwell-Boltzmannian. Fixing the distribution function of the relative velocities of the particles as Maxwell-Boltzmannian has serious physical implications for the nuclear reaction rate theory (Haubold and Mathai 1984). However, if we choose by the time the Maxwell-Bolzmannian approach to the nuclear reaction rate in (2.2) and (2.6) then the distribution of the relative velocities of the particles can be written in the following manner:

$$f(v)dv = \left(\frac{\mu}{2\pi kT}\right)^{\frac{3}{2}} \exp\left\{-\frac{\mu v^2}{2kT}\right\} 4\pi v^2 dv. \tag{2.9}$$

The function $f(v)$ satisfies the normalization condition $\int_0^\infty dv f(v) = 1$. As given in (2.9) we can take it as the Maxwell-Bolzmannian relative kinetic energy spectrum for a non-degenerate, non-relativistic gas of particles (cp. Fig. 2.4)

$$f(E)dE = \frac{2}{\pi^{1/2}(kT)^{3/2}} \exp\left\{-\frac{E}{kT}\right\} E^{\frac{1}{2}} dE. \tag{2.10}$$

For a detailed discussion of the underlying physical assumptions for the application of (2.10) or the evaluation of reaction rates see Haubold and Mathai (1984).

2.4 Nonresonant Neutron Capture Cross Section

In the case of a nuclear reaction via neutron capture the solution of the respective SCHRÖDINGER equation is the plane wave which can be normalized to unit particle density at infinity. Then the cross section is determined by the ratio of the square of the absolute value of the wave function, ψ, for inelastic scattering by the current density of the particles, **s**

$$\sigma(v) = \frac{|\psi|^2}{|\mathbf{s}|} = \frac{|\psi|^2}{|\rho \mathbf{v}|} \sim \frac{1}{v}, \tag{2.11}$$

where ρ denotes the space density and **v** the current velocity of the ψ-field, respectively. At low energy the s-wave interactions dominate. That is to say, the reactions take place chiefly through a particular angular momentum state $l = 0$. From (3.3) follows that σv is a constant. Derivations from this occur at higher energies when other partial waves become important and it is convenient to express the nonresonant slowly varying velocity dependence of the cross section,

$$\sigma(v) = \frac{T(v)}{v}, \tag{2.12}$$

as the first three terms of a MACLAURIN series in the relative velocity, v, of the neutron and the nucleus:

$$T(v) = T(0) + T'(0)v + \frac{1}{2}T''(0)v^2, \tag{2.13}$$

where $[T(0)] = cm^3 s^{-1}$, $[T'(0)] = cm^2$, and $[T''(0)] = cms$ (Fowler et al. 1967, 1975). In (2.13) $T(0)$, $T'(0)$, and $T''(0)$ are empirical constants measured in nuclear experiments, and the prime indicates differentiation with respect to v.

2.5 Nonresonant Charged Particle Cross Section

If two nuclei, of charges $Z_1 e$ and $Z_2 e$, and masses m_1 and m_2, collide with kinetic energy of relative motion, $E = \mu v^2/2$, then on the basis of non-relativistic quantum scattering theory one obtains, in spherical coordinates at $r = 0$, for the square of the wave function (Blatt and Weisskopf 1959):

$$|\psi(0)|^2 = |\Gamma(1 - i\eta(v))|^2 \exp(-\pi\eta(v)), \tag{2.14}$$

where

$$\eta(v) = \frac{Z_1 Z_2 e^2}{hv} = \frac{Z_1 Z_2 c\alpha}{v}; \tag{2.15}$$

$\eta(v)$ is the Sommerfeld parameter, h is the PLANCK quantum of action, α is the Sommerfeld fine structure constant, $\Gamma(\cdot)$ denotes the gamma function, and $i^2 = -1$. According to the elementary property of the gamma function we have from (2.14)

$$\begin{aligned}|\psi(0)|^2 &= \Gamma(1 + i\eta(v))\Gamma^*(1 + i\eta(v))\exp(-\pi\eta(v)) \\ &= \Gamma(1 + i\eta(v))\Gamma(1 - i\eta(v))\exp(-\pi\eta(v)) \\ &= \frac{\pi\eta(v)}{\sinh(\pi\eta(v))}\exp(-\pi\eta(v)) = \frac{2\pi\eta(v)}{\exp(2\pi\eta(v)) - 1}.\end{aligned} \tag{2.16}$$

For physical reasons the argument of the exponential function is always positive and thus the quantity in (2.16) is smaller than unity and for a strong COULOMB interaction, that means $2\pi\eta(v) \gg 1$, it becomes exponentially very small. This means that for strong repulsion and low energy collisions, the wave will only penetrate the potential barrier at the origin with a rapidly decreasing amplitude whose asymptotic form is

$$|\psi(0)|^2 \sim 2\pi\eta(v)\exp(-2\pi\eta(v)), \tag{2.17}$$

where the exponential factor is called Gamow factor and (2.17) is the barrier penetration factor. As in (2.11) the cross section for the charged particle reaction is obtained by dividing the absolute square value of the wave function for inelastic scattering (2.17) by the probability current density which is proportional to the velocity of the ψ-field. Then, we obtain the overall velocity dependence of the nuclear reaction cross section at low energies for nonresonant charged particle interactions by

$$\sigma(v) \sim \eta^2(v)\exp(-2\pi\eta(v)). \tag{2.18}$$

It is convenient to factor out the energy dependence and express the cross section, $\sigma(E)$, by

$$\sigma(E) = \frac{S(E)}{E} \exp(-2\pi\eta(E)), \qquad (2.19)$$

where

$$\eta(E) = \left(\frac{\mu}{2}\right)^{\frac{1}{2}} \frac{Z_1 Z_2 e^2}{h E^{1/2}}. \qquad (2.20)$$

Equation (2.19) defines the cross section factor, $S(E)$, representing the intrinsically nuclear parts of the probability for the occurrence of a nuclear reaction (cp. also Salpeter 1952). The cross section factor, $S(E)$, is often found to be constant or a slowly varying function of energy over a limited range of energy (Fowler et al. 1967, 1975). Far from a nuclear resonance $S(E)$ may be conveniently expressed in terms of the power series expansion,

$$S(E) = S(0) + S'(0)E + \frac{1}{2}S''(0)E^2, \qquad (2.21)$$

where $[S(0)] = MeV$ barns, $[S'(0)] =$ barns, and $[S''(0)] =$ barns MeV^{-1}. The prime indicates differentiation with respect to E.

2.6 Resonant Cross Section for Neutrons and Charged Particles

For a single resonance of energy, E_r, the cross section, $\sigma(E)$, of the nuclear reaction (2.1) can be represented as a function of energy in terms of the classical BREIT-WIGNER formula (Blatt and Weisskopf 1959),

$$\sigma(E) = \pi \lambda^2 \omega \frac{\Gamma_{12}\Gamma_{34}}{(E_r - E)^2 + (\Gamma/2)^2}, \qquad (2.22)$$

where $\lambda = h/(\mu v)$ is the reduced DE BROGLIE wavelength. The statistical factor ω is defined by $\omega = (2J+1)/[(2J_1+1)(2J_2+1)]$, where J is the angular momentum of the resonance state, and J_1 and J_2 are the angular momenta of particles 1 and 2, respectively. The total width, Γ, of the resonance state is given by $\Gamma = h/\tau = \Gamma_{12} + \Gamma_{34} + \cdots$, where τ is the effective lifetime of the state. The partial width, Γ_{12}, is the width for reemission of particles 1 and 2, and Γ_{34} is the width for emission of particles 3 and 4.

The partial width, Γ_{34}, for the absorption or emission of a certain particle by the compound nucleus, is a strong energy depending function and can be written (Blatt and Weisskopf 1959),

$$\Gamma_{34}(E) = \frac{2^{3/2}\mu^{1/2}R_0 D}{h} E^{1/2} P(E), \qquad (2.23)$$

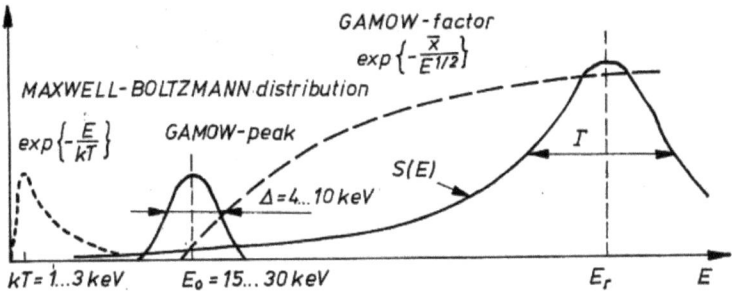

Fig. 2.3 Schematic plot of the energy-dependent factors for the integral of thermonuclear reaction rates: Maxwell-Boltzmann distribution function, nonresonant nuclear cross section, and resonant nuclear cross section

where R_0 is the characteristic wavelength of nucleons inside the nucleus (of the order of 10^{-13} cm), D is the average distance of levels, and $P(E)$ denotes the barrier penetration factor. At low energy the s-wave ($l = 0$) interactions dominate and the barrier penetration factor in (2.23) is given in (2.17), thus we have,

$$P(E) = 2\pi(\mu/2)^{\frac{1}{2}} \frac{Z_1 Z_2 e^2}{h E^{1/2}} \exp\left\{-2\pi(\mu/2)^{1/2} \frac{Z_1 Z_2 e^2}{h E^{1/2}}\right\}. \quad (2.24)$$

For the total width, Γ, we consider an ad hoc linear energy dependence of the form

$$\Gamma(E) = \Gamma_0 + \Gamma_1 E, \quad (2.25)$$

where Γ_0 and Γ_1 are empirical constants measured in nuclear experiments;.

Inserting (2.22)–(2.24) and (2.25) we obtain the parameterized form of the BREIT-WIGNER one-resonance-level formula

$$\sigma(E) = \left[\frac{\pi^2 h^2}{2\mu E}\right]\left[\frac{2^{3/2} \mu^{1/2} R_0 E^{1/2}}{h} \frac{2\pi \mu^{1/2} Z_1 Z_2 e^2}{2^{1/2} h E^{1/2}} \exp\left\{-\frac{2\pi \mu^{1/2} Z_1 Z_2 e^2}{2^{1/2} h E^{1/2}}\right\}\right]$$
$$\times \left[\frac{\omega \Gamma_{12} D}{(E_r - E)^2 + ([\Gamma_0 + \Gamma_1 E]/2)^2}\right]. \quad (2.26)$$

2.7 Parameterizations of Thermonuclear Reaction Rates

In the following, we consider the form of the quantity $<\sigma v>$ in (2.6) of Chap. 2, taking into account the distribution function of the relative particle velocities as given in (2.9) and (2.10), and the nuclear cross section derived in (2.12), (2.19) and (2.26), respectively.

Aside from the lorentz factor due to resonance phenomena in nuclear reactions, the two dominant factors in the quantity $<\sigma v>$ are the COULOMB barrier (Gamow factor), which inhibits the reaction rate at low energies, and the tail of the distribution function of relative velocities of the particles (Maxwell-Boltzmann factor) as shown in Fig. 2.3. Thus, the kernel of the integral of the quantity $<\sigma v>$ in (2.6) is a product of a rapidly rising cross section and a steeply falling distribution function, which gives a not quite asymmetrical peak, called the Gamow peak. For the treatment of nuclear reaction networks for cosmological or stellar nucleosynthetic calculations, these integrals have to be evaluated as far as possible in closed-form and the results represented in manageable analytic expressions (Fowler et al. 1967, 1975; Haubold and John 1982; Haubold and Mathai 1984).

To have a feeling for the integration problem of the quantity $<\sigma v>$ we refer to the schematic plot of typical charged particle and neutron cross sections, as a function of center-of-mass energy shown in Fig. 2.4 (Wagoner 1969).

Also, shown in Fig. 2.4 are two Maxwell-Boltzmann distribution functions of the relative particle velocities for temperatures near the limit of the region of interest (Wagoner 1969). Inserting the Maxwell distribution function (2.9) and (2.10), respectively, in (2.6), we obtain the following:

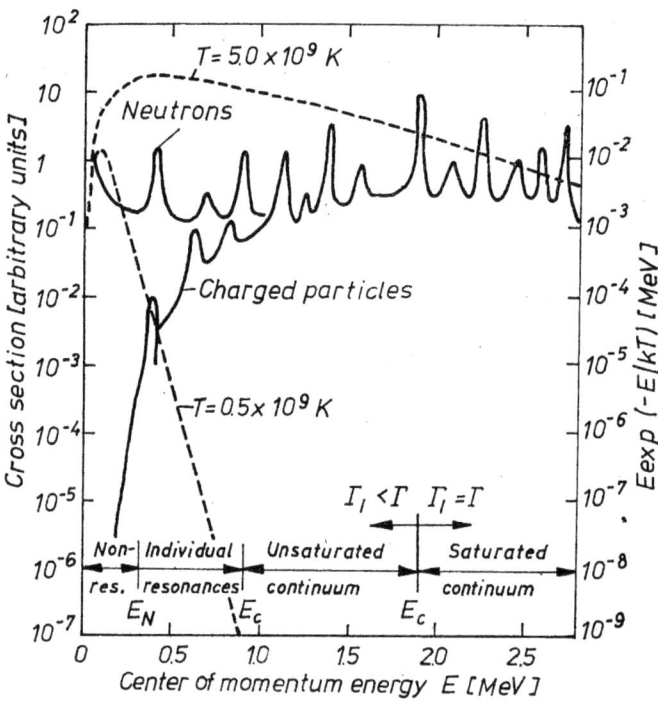

Fig. 2.4 Schematic plot of charged particle and neutron cross sections as a function of center-of-momentum energy, together with the Maxwell-Boltzmann distribution function (dashed lines) (Wagoner 1969)

2.7 Parameterizations of Thermonuclear Reaction Rates

$$<\sigma v> = \left(\frac{\mu}{2\pi kT}\right)^{\frac{3}{2}} \int_0^\infty dv\, \sigma(v) v^3 \exp\left\{-\frac{\mu v^2}{2kT}\right\}, \quad (2.27)$$

$$<\sigma v> = \left(\frac{8}{\pi\mu}\right)^{\frac{1}{2}} \frac{1}{(kT)^{\frac{3}{2}}} \int_0^\infty dE\, \sigma(E) E \exp\left\{-\frac{E}{kT}\right\}. \quad (2.28)$$

The remaining factor in the kernel of the integrand of (2.27) and (2.28), respectively, is the nuclear cross section which has quite special energy characteristics depending on whether the reaction proceeds via a resonant or nonresonant mechanism.

According to Wagoner (1969) (see also Fowler et al. 1967, 1975) it is convenient to divide the quantity $<\sigma v>$ in (2.27) and (2.28) into four parts.

$$<\sigma v> = <\sigma v>_{nr} + (\Sigma_{res} <\sigma v>_r) + <\sigma v>_{uc} + <\sigma v>_{sc}. \quad (2.29)$$

Splitting of $<\sigma v>$ in (2.29) in order of increasing energy as indicated in Fig. 2.4 corresponds to the successive synthesis of heavier elements out of lighter elements in stellar generations as discussed in Sect. 2.1 and outlined in Table 2.1. The nuclear reactions of the proton-proton chain are nonresonant (nr), the nuclear reactions of the CNO cycles are dominated by one or a few resonances (r), and the carbon burning, neon burning, oxygen burning, and silicon burning are characterized by strongly overlapping resonances (unsaturated continuum (uc) and saturated continuum (sc)).

As already indicated, the quantity $<\sigma v>$ depends strongly on the temperature. In a highly evolved star going through explosive nucleosynthesis, the thermal distribution of energies has a sufficiently broad spectrum that a large number of compound nuclear states contribute to the effective average cross section of a considered reaction. Thus, the high temperature regions of the quantity $<\sigma v>$, those are the last two terms on the right hand side of (4.3), must be treated by nuclear statistical models (e.g. HAUSER-FESHBACH model; c.f. Fowler 1974; Fowler et al. 1967, 1975). But, as shown by Fowler et al. (1967), (1975), even in the case of closely spaced or overlapping resonances, the formalism for thermonuclear reaction rates of nonresonant reactions, or reactions which proceed in the wing of a single resonance, still remains valid. At least, in all cases mentioned the cross section factor $S(E)$ is a smoothly varying function of energy as shown in Fig. 2.3. For that reason, in the following, we will concentrate our considerations on the closed-form evaluation of the first two terms for $<\sigma v>$ on the right hand side of (2.29). We shall consider the analytic representation of each of the first two terms on the right hand side in (2.29) separately.

2.8 Nonresonant Reaction Rates

Here, we consider various situations such as neutral particles, charged particles etc.

Case 2.1 The neutral particle case

Inserting (2.12) and (2.13) in (2.27) leads to the nonresonant neutron capture representation of $<\sigma v>$,

$$<\sigma v> = \frac{1}{\pi^{\frac{1}{2}}} \sum_{\nu=0}^{2} \frac{T^{(\nu)}(0)}{\nu!} 2^{\nu+1} \left(\frac{kT}{\mu}\right)^{\nu} \int_0^{\infty} dy\, y^{\nu+\frac{1}{2}} \exp\{-y\},$$

where $y = \mu v^2/(2kT)$, and we obtain,

$$<\sigma v> = \frac{1}{\pi^{\frac{1}{2}}} \sum_{\nu=0}^{2} \frac{T^{(\nu)}(0)}{\nu!} 2^{\nu+1} \left(\frac{kT}{\mu}\right)^{\nu} \Gamma\left(\nu+\frac{3}{2}\right), \Re(\nu) > -\frac{3}{2}. \quad (2.30)$$

With (2.30) and (2.2), the closed-form representation of the nonresonant thermonuclear reaction rate in the neutral particle case is obtained (cf. also Haubold and John 1982).

Case 2.2 The charged particle case with strict Maxwell-Boltzmann distribution function

Inserting (2.19) and (2.21) in (2.28) leads to the charged particle representation of $<\sigma v>$ if the reaction takes place far away from any resonance,

$$<\sigma v> = \left(\frac{8}{\pi \mu}\right)^{\frac{1}{2}} \frac{1}{(kT)^{\frac{3}{2}}} \sum_{\nu=0}^{2} \frac{S^{(\nu)}(0)}{\nu!} \int_0^{\infty} dE\, E^{\nu} \exp\left\{-\frac{E}{kT} - 2\pi\, \eta(E)\right\},$$

where the Sommerfeld parameter $\eta(E)$ is given in (2.20). With the substitution $y = E/(kT)$ we obtain

$$<\sigma v> = \left(\frac{8}{\pi\mu}\right)^{\frac{1}{2}} \sum_{\nu=0}^{2} \frac{S^{(\nu)}(0)}{\nu!} \frac{1}{(kT)^{-\nu+\frac{1}{2}}} \int_0^{\infty} dy\, y^{\nu} e^{-y-zy^{-\frac{1}{2}}}, \quad (2.31)$$

and remove the parameter-dependent integral

$$N_{i_\nu}(z) = \int_0^{\infty} dy\, y^{\nu} e^{-y-zy^{-\frac{1}{2}}} \quad (2.32)$$

from (2.31). With (2.32) we write for the quantity $<\sigma v>$ in (2.31)

2.8 Nonresonant Reaction Rates

$$<\sigma v> = \left(\frac{8}{\pi\mu}\right)^{\frac{1}{2}} \sum_{\nu=0}^{2} \frac{S^{(\nu)}(0)}{\nu!} \frac{1}{(kT)^{-\nu+\frac{1}{2}}} N_{i_\nu}(z), \tag{2.33}$$

where

$$z = 2\pi \left(\frac{\mu}{2kT}\right)^{\frac{1}{2}} \frac{Z_1 Z_2 e^2}{h}. \tag{2.34}$$

With the intention to include more general energy dependent nuclear cross section factors, $S(E)$, than given in (2.21), we considered the following general collision probability integral (Haubold and Mathai 1984):

$$N_1(z; a, \rho, n, m) = a \int_0^\infty dt\, t^{-n\rho} e^{-at - zt^{-\frac{n}{m}}}. \tag{2.35}$$

For the analytic evaluation of the integral $N_1(z; a, \rho, n, m)$ in (2.35), first we will give a general result and then get $N_{i_\nu}(z)$ in (2.32) as a special case. For a detailed description of the mathematical method to tackle integrals of the type in (2.35), see Haubold and Mathai (1984). The following theorem is originally due to Saxena (1960) and a simple proof of the theorem by using random variables is given by Haubold and Mathai (1984):

Theorem 2.1

$$N_1(z; a, \rho, n, m) = a \int_0^\infty dt\, t^{-n\rho} e^{-at - zt^{-\frac{n}{m}}}$$

$$= a^{n\rho} (2\pi)^{\frac{1}{2}(2-n-m)} m^{\frac{1}{2}} n^{\frac{1}{2} - n\rho}$$

$$\times G_{0,m+n}^{m+n,0}\left[\frac{z^m a^n}{m^m n^n} \Big|_{0, \frac{1}{m}, \ldots, \frac{m-1}{m}, \frac{1-n\rho}{n}, \ldots, \frac{n-n\rho}{n}}\right], \tag{2.36}$$

for $\Re(a) > 0$ and $\Re(z) > 0$; $G_{0,m+n}^{m+n,0}(\cdot)$ denotes Meijer's G-function (see Mathai and Saxena 1973).

Consider the case $m = 2, n = 1, a = 1$, and $\rho = -\nu$ in (2.36). Then, we have (cf. also Mathai 1971; Haubold and John 1982)

$$N_1(z; 1, -\nu, 1, 2) = \int_0^\infty dy\, y^\nu e^{-y - zy^{-\frac{1}{2}}} = \frac{1}{\pi^{\frac{1}{2}}} G_{0,3}^{3,0}\left[\frac{z^2}{4} \Big|_{a, \frac{1}{2}, 1+\nu}\right], \tag{2.37}$$

where z is given in (2.34). With (2.37) it is known that the nonresonant thermonuclear reaction rate can be represented in closed-form by using the highly efficient theory of generalized hypergeometric functions as given in Mathai and Saxena (1973). In the case of the Maxwell-Boltzmann approach to the nonresonant thermonuclear reaction rate Meijer's G-function of the type $G_{0,p}^{p,0}(\cdot)$ appears in the closed-form representation. For a critical review of the present status of the analytic evaluation of nonresonant thermonuclear reaction rates see Haubold and John (1982). Approximation consideration of $N_{i_\nu}(z)$ in (2.37) are given in papers of Bahcall (1966) and

Critchfield (1972). On the basis of the complex Mellin-Barnes integral representation of Meijer's G-function in (2.37), one can derive series representations of the integral $N_{i_\nu}(z)$ for all five classes of the parameters ν, $\nu \neq \pm\frac{1}{2}$, $\lambda = 0, 1, 2, \ldots$; ν a positive integer; ν a negative integer; ν a positive half integer; ν a negative half integer; as given in Sect. 4.3. (Haubold and Mathai 1984)). Moreover, one finds series representations of $N_{i_\nu}(z)$ which are termwise integrable over any finite range of the variable, z, given in (2.34) and which can be used for the numerical computation of $<\sigma v>$ in (2.33) also.

Case 2.3 The charged particle case with depleted Maxwell-Boltzmann distribution function

In the following, we admit a depletion of the high energy tail of the Maxwell-Boltzmann distribution function of the relative kinetic energies of the reacting particles given in (2.10). For the discussion of the physical reasons for a depletion of the high energy tail of the Maxwell-Boltzmann distribution function, we refer to the paper of Haubold and John (1982) and Haubold and Mathai (1986c).

For the integral of $<\sigma v>$ in comparison with the strict Maxwell-Boltzmannian case (2.32) we have now

$$N_{i_\nu}(z; \delta) = \int_0^\infty dy\, y^\nu e^{-y-y^\delta-zy^{-\frac{1}{2}}}. \tag{2.38}$$

Consider the general integral

$$N_2(z; a, b, \delta, \nu, n, m) = \int_0^\infty dt\, t^\nu e^{-at-bt^\delta-zt^{-\frac{n}{m}}},$$

where $z > 0, a > 0, b > 0, m, n = 1, 2, 3, \ldots$. Since

$$e^{-bt^\delta} = \sum_{k=0}^\infty \frac{(-b)^k}{k!} t^{k\delta},$$

we have

$$N_2(z; a, b, \delta, \nu, n, m) = \sum_{k=0}^\infty \frac{(-b)^k}{k!} \int_0^\infty dt\, t^{\nu+k\delta} e^{-at-zt^{-\frac{n}{m}}}. \tag{2.39}$$

According to the Theorem 2.1 (cf. also Haubold and Mathai 1984) we have,

$$N_2(z; a, b, \delta, \nu, n, m) = \sum_{k=0}^\infty \frac{(-b)^k}{k!} a^{-(\nu+k\delta+1)} (2\pi)^{\frac{1}{2}(2-n-m)} m^{\frac{1}{2}} n^{\frac{1}{2}+\nu+k\delta}$$

$$\times G_{0,m+n}^{m+n,0}\left[\frac{z^m a^n}{m^m n^n} \bigg|_{0, \frac{1}{m}, \ldots, \frac{m-1}{m}, \frac{1+\nu+k\delta}{n}, \ldots, \frac{n+\nu+k\delta}{n}}\right].$$

2.8 Nonresonant Reaction Rates

Now, we consider (2.38) and have by putting $a = 1, b = 1, n = 1, m = 2$

$$N_2(z; 1, 1, \delta, \nu, 1, 2) = \frac{1}{\pi^{\frac{1}{2}}} \sum_{k=0}^{\infty} \frac{(-1)^k}{k!} G_{0,3}^{3,0}\left[\frac{z^2}{4}\Big|_{0,\frac{1}{2},1+\nu+k\delta}\right]. \tag{2.40}$$

For the numerical computation of (2.40) all the cases in Sect. 4.3 of Haubold and Mathai (1984) can apply with ν replaced by $\nu + k\delta, k = 0, 1, 2, \ldots$ Note that if δ is an irrational number then $\nu + k\delta$ cannot be a negative integer or negative half integer or zero or positive integer or positive half integer unless ν is also a suitable irrational number. Hence, in this case the poles of the integrand will be simple and the G-function can be easily evaluated as in the Case (i) of Sect. 4.3 of Haubold and Mathai (1984).

Case 2.4 The charged particle case with modified Maxwell-Boltzmann distribution function

If there occurs a cut off of the high energy tail of the Maxwell-Boltzmann distribution function given in (2.10) we expect that the closed-form representation of the appropriate quantity $<\sigma v>$, given by

$$N_{i_\nu}(z; d) = \int_0^d dy \, y^\nu e^{-y-zy^{-\frac{1}{2}}}, \tag{2.41}$$

will not lead to the type $G_{0,p}^{p,0}(\cdot)$ of Meijer's G-function as we have had in the cases in (2.37) and (2.40), respectively. For the discussion of physical reasons for the cut off modification of the Maxwell-Boltzmann distribution function of the relative kinetic energy of the reacting particles, we refer to the papers of Haubold and John (1982) and Haubold and Mathai (1986a).

Again, we consider a more general form of the integral in (2.41) as

$$N_3(z; d, a, \rho, n, m) = \int_0^d dt \, t^{-n\rho} e^{-at - zt^{-\frac{n}{m}}}, \tag{2.42}$$

where $a > 0, d > 0, z > 0$. This can be evaluated by working out the density of a product of two independent real random variables by using two different techniques. For the detailed discussion of the mathematical methods to tackle integrals of the types as given in (2.42), we refer to Haubold and Mathai (1984, 1986a). The results contain the following theorem:

Theorem 2.2 *For $z > 0, d > 0, a > 0, m$ and n positive integers,*

$$N_3(z; d, a, \rho, n, m) = \int_0^d dt \, t^{-n\rho} e^{-at - zt^{-\frac{n}{m}}} = \frac{m^{\frac{1}{2}}}{n}(2\pi)^{\frac{1-m}{2}} d^{-n\rho+1}$$

$$\times \sum_{r=0}^{\infty} \frac{(-ad)^r}{r!} G_{n,m+n}^{m+n,0}\left[\frac{z^m}{d^n m^m}\Big|_{-\rho+\frac{r+1}{n}+\frac{j-1}{n}, j=1,\ldots,n, \frac{i-1}{m}, j=1,\ldots,m}^{-\rho+\frac{r+2}{n}+\frac{j-1}{n}, j=1,\ldots,n}\right]$$

To obtain the special case realized in (2.41) we put $n = 1, m = 2, a = 1, \rho = \nu$, we get

$$N_3(z; d, 1, \nu, 1, 2) = \frac{1}{\pi^{\frac{1}{2}}} d^{-\nu+1} \sum_{r=0}^{\infty} \frac{(-d)^r}{r!} G_{1,3}^{3,0} \left[\frac{z^2}{4d} \Big|_{-\nu+r+1, 0, \frac{1}{2}}^{-\nu+r+2} \right]. \qquad (2.43)$$

For computable series representation of $N_{i_\nu}(z; d)$ in (2.43) for all classes of the parameters $-\nu + r + 1$ see Sect. 4.4 of Haubold and Mathai (1986a).

2.9 Resonant Reaction Rates

Here also we will consider various situations.

Case 2.5 The charged particle case with Maxwell-Boltzmann distribution function

We put (2.26) into (2.28) and obtain the representation of $<\sigma v>$ for resonant thermonuclear reactions (2.1):

$$<\sigma v> = (2\pi)^{\frac{5}{2}} \frac{Z_1 Z_2 e^2 R_0 \omega \Gamma_{12} D}{\mu^{\frac{1}{2}} (kT)^{\frac{3}{2}}} \int_0^\infty dE \exp\left\{ -\frac{E}{kT} - \frac{\bar{q}}{E^{\frac{1}{2}}} \right\}$$

$$\times \left[(E_r - E)^2 + \left(\frac{1}{2}(\Gamma_0 + \Gamma_1 E_1) \right)^2 \right]^{-1} \qquad (2.44)$$

where $\bar{q}$ is given by

$$\bar{q} = 2\pi \left(\frac{\mu}{2} \right)^{\frac{1}{2}} \frac{Z_1 Z_2 e^2}{h} = z(kT)^{\frac{1}{2}}, \qquad (2.45)$$

where z is given in (2.34). From (2.44) we remove the integral

$$R = \int_0^\infty dE \exp\left\{ -\frac{E}{kT} - \frac{\bar{q}}{E^{\frac{1}{2}}} \right\} \left[(E_r - E)^2 + \left(\frac{1}{2}(\Gamma_0 + \Gamma_1 E) \right)^2 \right]^{-1} \qquad (2.46)$$

which may be written more conveniently as,

$$R = \left[1 + \left(\frac{\Gamma_1}{2} \right)^2 \right]^{-1} \int_0^\infty dE \exp\left\{ -\frac{E}{kT} - \frac{\bar{q}}{E^{\frac{1}{2}}} \right\} \left[(\tilde{E}_r - E)^2 + \left(\frac{\tilde{\Gamma}}{2} \right)^2 \right]^{-1} \qquad (2.47)$$

where $\tilde{E}_r$ denotes a modified resonance energy,

2.9 Resonant Reaction Rates

$$\tilde{E}_r = \left[E_r - \frac{\Gamma_0 \Gamma_1}{4}\right]\left[1 + \left(\frac{\Gamma_1}{2}\right)^2\right]^{-1}, \tag{2.48}$$

with $\tilde{\Gamma}$ as the modified total width,

$$\tilde{\Gamma} = [\Gamma_0 + E_r \Gamma_1]\left[1 + \left(\frac{\Gamma_1}{2}\right)^2\right]^{-1}. \tag{2.49}$$

The form of the resonance denominator with Γ_0 is conserved if one transforms $\Gamma_0 \to \Gamma_0 + E\Gamma_1$. Choosing the variable $E = y[1 + (\frac{\Gamma_1}{2})^2]^{-1}$ leads to

$$R = \int_0^\infty dy \exp\left\{-y[kT(1+(\Gamma_1/2)^2)]^{-1} - y^{-\frac{1}{2}}\bar{q}(1+(\Gamma_1/2)^2)^{\frac{1}{2}}\right\}$$
$$\times [(E_r - y - \Gamma_0\Gamma_1/4)^2 + ((\Gamma_0 + E_r\Gamma_1)/2)^2]^{-1}.$$

With the notations

$$a = [kT(1+(\Gamma_1/2)^2)]^{-1}, \quad b = E_r - \frac{1}{4}\Gamma_0\Gamma_1,$$

$$g = \frac{1}{2}(\Gamma_0 + E_r\Gamma_1), \quad q = \bar{q}(1+(\Gamma_1/2)^2)^{\frac{1}{2}},$$

we obtain

$$R(q,a,b,g) = \int_0^\infty dy \frac{\exp\{-ay - qy^{-\frac{1}{2}}\}}{(b-y)^2 + g^2}$$

and (2.44) can be written as,

$$<\sigma v> = (2\pi)^{\frac{5}{2}} \frac{Z_1 Z_2 e^2 R_0 \omega \Gamma_{12} D}{\mu^{\frac{1}{2}}(kT)^{\frac{3}{2}}}[1+(\Gamma_1/2)^2]^{-1} R(q,a,b.g)$$

(see Haubold and John 1979; Haubold and Mathai 1986b).

In order to comprehend cases in which the cross section factor or the partial width are additionally multiplied by energy-depending factors we consider the more general integral

$$R_1(q,a,b,g;\nu,n,m) = \int_0^\infty dt \frac{\exp\{-at - qt^{-\frac{n}{m}}\}}{(b-t)^2 + g^2}. \tag{2.50}$$

We may replace the denominator $[(b-t)^2 + g^2]^{-1}$ in (2.50) by an equivalent integral for $g^2 > 0$. That is,

$$\frac{1}{(b-t)^2 + g^2} = \int_0^\infty dx \, \exp\{-[(b-t)^2 + g^2]x\}. \tag{2.51}$$

But, we can write

$$e^{-x(b-t)^2} = \sum_{k=0}^{\infty} \frac{(-1)^k}{k!} x^k (b-t)^{2k}$$

$$= \sum_{k=0}^{\infty} \frac{(-1)^k}{k!} x^k \sum_{k_1=0}^{2k} \binom{2k}{k_1} (-1)^{k_1} b^{2k-k_1} t^{k_1}, \quad (2.52)$$

where, for example, $\binom{m}{n} = \frac{m!}{n!(m-n)!}$, $0! = 1$. From (2.50) and (2.51) and (2.52) we have,

$$R_1(q,a,b,g;\nu,n,m) = \int_0^\infty dx\, e^{-g^2 x} \sum_{k=0}^{\infty} \frac{(-1)^k}{k!} x^k \sum_{k_1=0}^{2k} \binom{2k}{k_1} (-1)^{k_1} b^{2k-k_1}$$

$$\times \int_0^\infty dt\, t^{\nu+k_1} e^{-at - qt^{-\frac{n}{m}}}. \quad (2.53)$$

Note that

$$\int_0^\infty dx\, x^k e^{-g^2 x} = \int_0^\infty dx\, x^{(k+1)-1} e^{-g^2 x} = \Gamma(k+1)(g^2)^{-(k+1)} = \frac{k!}{g^2 (g^2)^k}, \quad (2.54)$$

and according to Theorem 2.1, we have

$$\int_0^\infty dt\, t^{\nu+k_1} e^{-at - qt^{-\frac{n}{m}}} = a^{-(\nu+k_1+1)} (2\pi)^{\frac{1}{2}(2-n-m)} m^{\frac{1}{2}} n^{\frac{1}{2}+\nu+k_1}$$

$$\times G_{0,m+n}^{m+n,0}\left[\frac{q^m a^n}{m^m n^n}\bigg|_{0,\frac{1}{m},\ldots,\frac{m-1}{m},\frac{1+\nu+k_1}{n},\ldots,\frac{n+\nu+k_1}{n}}\right] \quad (2.55)$$

(cp. also Mathai 1971; Saxena 1960; Haubold and Mathai 1984). Substituting (2.54) and (2.55) in (2.53) one has the following:

$$R_1(q,a,b,g;\nu,n,m) = \sum_{k=0}^{\infty} \frac{(-1)^k}{g^2(g^2)^k} \sum_{k_1=0}^{2k} \binom{2k}{k_1} (-1)^{k_1} b^{2k-k_1}$$

$$\times a^{-(\nu+k_1+1)} (2\pi)^{\frac{1}{2}(2-n-m)} m^{\frac{1}{2}} n^{\frac{1}{2}+\nu+k_1}$$

$$\times G_{0,m+n}^{m+n,0}\left[\frac{q^m a^n}{m^m n^n}\bigg|_{0,\frac{1}{m},\ldots,\frac{m-1}{m},\frac{1+\nu+k_1}{n},\ldots,\frac{n+\nu+k_1}{n}}\right]. \quad (2.56)$$

Put $n = 1$, $m = 2$, $\nu = 0$, to get $R(q,a,b,g)$.

2.9 Resonant Reaction Rates

$$R_1(q, a, b, g; 0, 1, 2) = R(q, a, b, g) = \frac{1}{g^2 a \pi^{\frac{1}{2}}} \sum_{k=0}^{\infty} \frac{(-1)^k}{(g^2)^k}$$

$$\times \sum_{k_1=0}^{2k} \binom{2k}{k_1} \frac{(-1)^{k_1}}{a^{k_1}} b^{2k-k_1} G_{0,3}^{3,0} \left[\frac{q^2 a}{4} \Big|_{0, \frac{1}{2}, 1+k_1} \right] \quad (2.57)$$

for $(b - v/a)^2/g^2 < 1$, where $v = (q^2 a/4)^{\frac{1}{3}}$ (Haubold and Mathai 1986b).

Comparing the resonant result (2.57) with the nonresonant case (2.37) we observe that the former is an infinite sum over nonresonant contributions (Note that $q^2 a = z^2$). The appearance of the G-function of the type $G_{0,p}^{p,0}(\cdot)$ in (2.57) is due to the Maxwell-Boltzmannian approach to the resonant thermonuclear reaction rate. For the numerical computation of (2.57) all results obtained in Sect. 4.3 (Haubold and Mathai 1984) are applicable.

Case 2.6 The neutral particle case with Maxwell-Boltzmann distribution function

Considering neutron reactions, i.e., $q = 0$, we are dealing with energy-independent partial width. In the case $q = 0$, this also means $\Gamma_1 = 0$, for (2.53) we obtain

$$R_1(0, \bar{a}, \bar{b}, \bar{g}; 0, 0, 0) = \int_0^\infty dx \, e^{-\bar{g}^2 x} \sum_{k=0}^{\infty} \frac{(-1)^k}{k!} x^k$$

$$\times \sum_{k_1=0}^{2k} \binom{2k}{k_1} (-1)^{k_1} \bar{b}^{2k-k_1} \int_0^\infty dt \, t^{\nu+k_1} e^{-\bar{a}t}, \quad (2.58)$$

where

$$\bar{a} = \frac{1}{kT}, \quad \bar{b} = E_r, \quad \bar{g} = \frac{1}{2} \Gamma_0. \quad (2.59)$$

Taking into account (2.53) we get for (2.58)

$$R_1(0, \bar{a}, \bar{b}, \bar{g}; 0, 0, 0) = \sum_{k=0}^{\infty} \frac{(-1)^k}{\bar{g}^2 (\bar{g}^2)^k} \sum_{k_1=0}^{2k} \binom{2k}{k_1} (-1)^{k_1} \bar{b}^{2k-k_1} \frac{(\nu+k_1)!}{\bar{a}(\bar{a})^{\nu+k_1}}. \quad (2.60)$$

With (2.60), the closed-form representation of the resonant thermonuclear reaction rate for neutron reactions is obtained.

Case 2.7 The charged particle case with modified Maxwell-Boltzmann distribution function

In the following, we consider a modification of the Maxwell-Boltzmann distribution for resonant nuclear reactions. Instead of (2.50) we write

$$R_2(q, a, b, c, g; \nu, \delta, n, m) = \int_0^\infty dt \, t^\nu \frac{\exp\{-at - at^\delta - qt^{-\frac{n}{m}}\}}{(b-t)^2 + g^2}. \quad (2.61)$$

Replacing the denominator of the kernel of the integral (2.61) as in the case (2.51) we obtain the following, denoting the sequence of parameters $q, a, b, c, g; \nu, \delta, n, m$ by W, that is, $W = \{q, a, b, c, g; \nu, \delta, n, m\}$:

$$R_2(W) = \int_0^\infty dx \, e^{-g^2 x} \int_0^\infty dt \, t^\nu e^{-at - qt^{-\frac{n}{m}} - ct^\delta - x(b-t)^2}. \qquad (2.62)$$

But

$$e^{-(ct^\delta + x(b-t)^2)} = \sum_{k=0}^\infty (-1)^k \frac{[ct^\delta + x(b-t)^2]^k}{k!}$$

$$= \sum_{k=0}^\infty \frac{(-1)^k}{k!} \sum_{k_1=0}^k \binom{k}{k_1} c^{k-k_1} t^{\delta(k-k_1)} x^{k_1} (b-t)^{2k_1}$$

$$= \sum_{k=0}^\infty \frac{(-1)^k}{k!} \sum_{k_1=0}^k \binom{k}{k_1} c^{k-k_1} x^{k_1} \sum_{k_2=0}^{2k_1} \binom{2k_1}{k_2}$$

$$\times b^{2k_1 - k_2} (-1)^{k_2} t^{k_2 + \delta(k-k_1)}. \qquad (2.63)$$

Hence,

$$R_2(W) = \sum_{k=0}^\infty \frac{(-1)^k}{k!} \sum_{k_1=0}^k \binom{k}{k_1} c^{k-k_1} \sum_{k_2=0}^{2k_1} \binom{2k_1}{k_2} b^{2k_1-k_2} (-1)^{k_2}$$

$$\times \int_0^\infty dx \, x^{k_1} e^{-g^2 x} \int_0^\infty dt \, t^{\nu + k_2 + \delta(k - k_1)} e^{-at - qt^{-\frac{n}{m}}}, \qquad (2.64)$$

and due to the Theorem 2.1 we have

$$R_2(W) = \sum_{k=0}^\infty \frac{(-1)^k}{k!} \sum_{k_1=0}^k \binom{k}{k_1} g^{-2(k_1+1)} k_1! c^{k-k_1} \sum_{k_2=0}^{2k_1} \binom{2k_1}{k_2} b^{2k_1-k_2} (-1)^{k_2}$$

$$\times a^{-(\nu + k_2 + \delta(k-k_1) + 1)} m^{\frac{1}{2}} n^{\nu + k_2 + \delta(k-k_1) + \frac{1}{2}}$$

$$\times G_{0, m+n}^{m+n, 0} \left[\frac{q^m a^n}{m^m n^n} \Big|_{0, \frac{1}{m}, \dots, \frac{m-1}{m}, \frac{1 + \nu + k_2 + \delta(k-k_1)}{n}, \dots, \frac{n + \nu + k_2 + \delta(k-k_1)}{n}} \right].$$

That is,

2.10 Series Representations for the Thermonuclear Functions: $G_{0,3}^{3,0}(\cdot)$ 37

$$R_2(W) = (2\pi)^{\frac{1}{2}(2-n-m)} m^{\frac{1}{2}} n^{\nu+\frac{1}{2}} a^{-(\nu+1)} g^{-1} \sum_{k=0}^{\infty} \frac{(-1)^k}{k!} \sum_{k_1=0}^{k} \binom{k}{k_1} g^{-k_1} k_1! c^{k-k_1}$$

$$\times \sum_{k_2=0}^{2k_1} \binom{2k_1}{k_2} b^{2k_1-k_2} (-1)^{k_2} a^{-(k_2+\delta(k-k_1))} n^{k_2+\delta(k-k_1)}$$

$$\times G_{0,m+n}^{m+n,0}\left[\frac{q^m a^n}{m^m n^n}\Big|_{0,\frac{1}{m},\dots,\frac{m-1}{m},\frac{1+\nu+k_2+\delta(k-k_1)}{n},\dots,\frac{n+\nu+k_2+\delta(k-k_1)}{n}}\right]. \quad (2.65)$$

for $|(b - \frac{n}{a} u^{\frac{1}{m+n}})^2/g^2| < 1$, where $u = q^m a^n/(m^m n^n)$. Putting $n = 1, m = 2, \nu = 0$ we obtain

$$R_2(q, a, b, c, g; 0, \delta, 1, 2) = \frac{1}{\pi^{\frac{1}{2}} ag} \sum_{k=0}^{\infty} \frac{(-1)^k}{k!} \sum_{k_1=0}^{k} \binom{k}{k_1} g^{-k_1} k_1! c^{k-k_1}$$

$$\times \sum_{k_2=0}^{2k_1} \binom{2k_1}{k_2} b^{2k_1-k_2} (-1)^{k_2} a^{-(k_2+\delta(k-k_1))}$$

$$\times G_{0,3}^{3,0}\left[\frac{q^2 a}{4}\Big|_{0,\frac{1}{2},1+2+\delta(k-k_1)}\right], \quad (2.66)$$

for $|(b - (q/2a)^{\frac{4}{3}})/g^2| < 1$. Series representations for the numerical computation of the G-function contained in (2.66) are given in Sect. 2.10.

2.10 Series Representations for the Thermonuclear Functions: $G_{0,3}^{3,0}(\cdot)$

Now, we are going back to the collision probability integral containing as a part of the kernel a Maxwell-Boltzmann distribution function term. Consider the case $m = 2, n = 1, a = 1$, and $\rho = -\nu$ in Eq. (2.35). Then, the closed-form representation of the thermonuclear reaction rate integral contains the Meijer's G-function of the following type: see for example (2.37),

$$\pi^{-\frac{1}{2}} G_{0,3}^{3,0}\left[\frac{z^2}{4}\Big|_{0,\frac{1}{2},1+\nu}\right]. \quad (2.67)$$

In the following, we shall derive representations of (2.67) which will be suitable for the numerical evaluation. In the light of the results obtained by Critchfield (1972) we refer to (2.67) as the thermonuclear functions occurring in Eqs. (2.37), (2.40), (2.57) and (2.66), respectively.

Case 2.8 $\nu \neq \pm\frac{\lambda}{2}, \lambda = 0, 1, 2, \ldots$

Then the poles of the Mellin-Barnes integral representation of (2.67) are simple and then the G-function has a simple series expansion. Consider the evaluation of the G-function in this case (Mathai and Saxena 1973):

$$G_{0,3}^{3,0}\left[\frac{z^2}{4}\Big|_{0,\frac{1}{2},1+\nu}\right] = \frac{1}{2\pi i}\int ds\, \Gamma(s)\Gamma\left(\frac{1}{2}+s\right)\Gamma(1+\nu+s)\left(\frac{z^2}{4}\right)^{-s}.$$

The poles are at $s = 0, -1, -2, \ldots; s = -\frac{1}{2}, \frac{1}{2} - 1, \ldots; s = -1-\nu, -1-\nu-1, \ldots$. The sum of the residues corresponding to the poles $s = 0, -1, -2, \ldots$, denoted by S_1, is the following:

$$S_1 = \sum_{r=0}^{\infty} \frac{(-1)^r}{r!}\Gamma\left(\frac{1}{2}-r\right)\Gamma(1+\nu-r)\left(\frac{z^2}{4}\right)^r$$

$$= \Gamma\left(\frac{1}{2}\right)\Gamma(1+\nu)\sum_{r=0}^{\infty}\frac{(-1)^r}{r!}\frac{1}{(1/2)_r(-\nu)_r}\left(\frac{z^2}{4}\right)^r$$

$$= \Gamma\left(\frac{1}{2}\right)\Gamma(1+\nu){}_0F_2\left(-;\frac{1}{2},-\nu;-\frac{z^2}{4}\right), \qquad (2.68)$$

where ${}_pF_q(\cdot)$ denotes the generalized hypergeometric function. The sum of the residues corresponding to the poles $s = -\frac{1}{2}, -\frac{1}{2} - 1, \ldots$, denoted by S_2, is the following:

$$S_2 = \sum_{r=0}^{\infty}\frac{(-1)^r}{r!}\Gamma\left(-\frac{1}{2}-r\right)\Gamma\left(\frac{1}{2}+\nu-r\right)\left(\frac{z^2}{4}\right)^{\frac{1}{2}+r}$$

$$= \Gamma\left(-\frac{1}{2}\right)\Gamma\left(\frac{1}{2}+\nu\right)\left(\frac{z^2}{4}\right)^{\frac{1}{2}}{}_0F_2\left(-;\frac{3}{2},\frac{1}{2}-\nu;-\frac{z^2}{4}\right). \qquad (2.69)$$

Further on the sum of the residues corresponding to the poles $s = -1-\nu, -1-\nu-1, \ldots$, denoted by S_3, is the following:

$$S_3 = \sum_{r=0}^{\infty}\frac{(-1)^r}{r!}\Gamma(-1-\nu-r)\Gamma\left(-\frac{1}{2}-\nu-r\right)\left(\frac{z^2}{4}\right)^{1+\nu+r}$$

$$= \Gamma(-1-\nu)\Gamma\left(-\frac{1}{2}-\nu\right)\left(\frac{z^2}{4}\right)^{1+\nu}{}_0F_2\left(-;\nu+2,\nu+\frac{3}{2};-\frac{z^2}{4}\right). \qquad (2.70)$$

Now we have the following result:

Theorem 2.3 *From S_1, S_2, and S_3 of (2.68),(2.69) and (2.70), respectively, we have*

$$G_{0,3}^{3,0}\left[\frac{z^2}{4}\Big|_{0,\frac{1}{2},1+\nu}\right] = S_1 + S_2 + S_3$$

for $\nu \neq \pm\frac{\lambda}{2}, \lambda = 0, 1, 2, \ldots$.

2.10 Series Representations for the Thermonuclear Functions: $G_{0,3}^{3,0}(\cdot)$

Note that the series S_1, S_2 and S_3 are termwise integrable over any finite range and hence computations can be carried out by using these ${}_0F_2(\cdot)$ functions (cf. Mathai and Saxena 1973).

Case 2.9 ν a positive integer

In this case the poles of the gammas in the integral representation of (2.67) are the following: The poles of $\Gamma(s)$ are at $s = 0, -1, -2, \ldots, -\nu, -\nu-1, \ldots$. The poles of $\Gamma(1+\nu+s)$ are at $s = -\nu-1, -\nu-2, \ldots$. The poles of $\Gamma\left(\frac{1}{2}+s\right)$ are at $s = -\frac{1}{2}, -\frac{1}{2}-1, \ldots$. Note that the poles at $s = -\frac{1}{2}, -\frac{1}{2}-1, \ldots$ and at $s = 0, -1, \ldots, -\nu$ are of order one each and the poles at $s = -1-\nu, -1-\nu-1, \ldots$ are of order two each.

Again, for convenience, the sum of the residues will be denoted by S_j, $j = 1, 2, 3$. The sum of the residues corresponding to the poles $s = -\frac{1}{2}, -\frac{1}{2}-1, \ldots$, denoted by S_1, is the following, by following through the same procedure as in the earlier case:

$$S_1 = \Gamma\left(-\frac{1}{2}\right)\Gamma\left(\frac{1}{2}+\nu\right)\left(\frac{z^2}{4}\right)^{\frac{1}{2}}{}_0F_2\left(-;\frac{3}{2},\frac{1}{2}-\nu;-\frac{z^2}{4}\right). \tag{2.71}$$

For the sum of the residues corresponding to the poles at $s = 0, -1, \ldots, -\nu$, denoted by S_2, we get the following:

$$S_2 = \Gamma(1+\nu)\Gamma\left(\frac{1}{2}\right)\sum_{r=0}^{\nu}\frac{1}{(1/2)_r(-\nu)_r r!}\left(\frac{z^2}{4}\right)^r. \tag{2.72}$$

For poles of order two, we will have to use the general formula for evaluating the residues when the poles are of higher orders, and in the present case, the integrand has to be differentiated once. Let the integrand of the G-function in (2.67) be denoted by $\Delta(s)$. That is,

$$\Delta(s) = \Gamma(s)\Gamma(1+\nu+s)\Gamma\left(\frac{1}{2}+s\right)\left(\frac{z^2}{4}\right)^{-s}.$$

Then, the sum of the residues corresponding to the poles $s = -\nu-1, \nu-2, \ldots$, we have the following, denoted by S_3:

$$S_3 = \sum_{r=0}^{\infty}\lim_{s\to -1-\nu-r}\frac{\partial}{\partial s}\{(s+1+\nu+r)^2\Delta(s)\}$$

$$= \sum_{r=0}^{\infty}\left(\frac{z^2}{4}\right)^{1+\nu+r}\lim_{s\to -1-\nu-r}\left\{-\ln\left(\frac{z^2}{4}\right)+\frac{\partial}{\partial s}\right\}B(s)$$

$$B(s) = \{(s+1+\nu+r)^2\Gamma(1+\nu+s)\Gamma(s)\Gamma(s+1/2)\}.$$

But, by inserting the factors $(s + 1 + \nu + r - 1)^2 \cdots (s + 1 + \nu)^2(s + \nu)(s + \nu - 1) \cdots s$ in the numerator and in the denominator, we can write $B(s)$ as the following:

$$B(s) = \frac{(s+1+\nu+r)^2(s+\nu+r)^2 \cdots (s+1+\nu)^2(s+\nu) \cdots s}{(s+\nu+r)^2 \cdots (s+1+\nu)^2(s+\nu) \cdots s}$$
$$\times \Gamma(1+\nu+r+s)\Gamma(s)\Gamma\left(\frac{1}{2}+s\right)$$
$$= \frac{\Gamma^2(s+r+\nu+2)\Gamma\left(\frac{1}{2}+s\right)}{(s+r+\nu)^2 \cdots (s+\nu+1)^2(s+\nu) \cdots s}$$

Now, by taking the limit we have the following, denoting the resulting quantity as B_r:

$$B_r = \lim_{s \to -1-\nu-r} B(s) = \frac{\Gamma^2(1)\Gamma\left(\frac{1}{2} - 1 - \nu - r\right)}{(-1)^2 \cdots (-r)^2(-r-1) \cdots (-r-\nu-1)}$$
$$= \frac{(-1)^{1+\nu+2r}\Gamma^2(1)\Gamma\left(-\frac{1}{2} - \nu - r\right)}{r!(r+\nu+1)!} = \frac{(-1)^{1+\nu+r}\Gamma\left(-\frac{1}{2} - \nu\right)}{r!(\nu+r+1)!\left(\frac{3}{2}+\nu\right)_r}. \quad (2.73)$$

Let

$$A(s) = \frac{\partial}{\partial s}\ln B(s)$$
$$= 2\psi(s+r+\nu+2) + \psi\left(\frac{1}{2}+s\right) - \frac{2}{s+r+\nu} - \frac{2}{s+r+\nu-1} - \cdots$$
$$- \frac{2}{s+\nu+1} - \frac{1}{s+\nu} - \cdots - \frac{1}{s}.$$

Consider the limit and let

$$A_r = \lim_{s \to -1-r-\nu} A(s)$$
$$= 2\psi(1) + \psi\left(-\frac{1}{2} - \nu - r\right) + \frac{2}{1} + \frac{2}{2} + \cdots + \frac{2}{r} + \frac{1}{r+1} + \cdots + \frac{1}{r+\nu+1}$$
$$= \psi(r+1) + \psi(r+\nu+2) + \psi\left(-\frac{1}{2} - \nu - r\right), \quad (2.74)$$

where $\psi(z)$ is the psi function or digamma function (cf. Mathai and Saxena 1973). Hence, for the sum of the residues corresponding to the poles $s = -1 - \nu, -1 - \nu - 1, \ldots$, we obtain

$$S_3 = \sum_{r=0}^{\infty} \left(\frac{z^2}{4}\right)^{1+\nu+r} \left\{-\ln\left(\frac{z^2}{4}\right) + A_r\right\} B_r, \quad (2.75)$$

2.10 Series Representations for the Thermonuclear Functions: $G_{0,3}^{3,0}(\cdot)$

where A_r and B_r are defined in (2.74) and (2.73) respectively. Hence, we have the following result:

Theorem 2.4 *For ν a positive integer,*

$$\pi^{-\frac{1}{2}} G_{0,3}^{3,0}\left[\frac{z^2}{4}\bigg|_{0,\frac{1}{2},1+\nu}\right] = \Gamma(1+\nu) \sum_{r=0}^{\nu} \frac{1}{(1/2)_r (-\nu)_r r!} \left(-\frac{z^2}{4}\right)^r$$

$$- 2\Gamma\left(\frac{1}{2}+\nu\right) \left(\frac{z^2}{4}\right)^{\frac{1}{2}} {}_0F_2\left(-; \frac{3}{2}, \frac{1}{2}-\nu; -\frac{z^2}{4}\right)$$

$$+ \pi^{-\frac{1}{2}} \left(\frac{z^2}{4}\right)^{1+\nu} \sum_{r=0}^{\infty} \left(\frac{z^2}{4}\right)^r \left\{-\ln\left(\frac{z^2}{4}\right) + A_r\right\} B_r,$$

where B_r and A_r are defined in (2.73) and (2.74) respectively.

Case 2.10 ν a negative integer

Let $\nu = -\mu$, $\mu = 2, 3, \ldots$. In this case, the poles of the G-function in (4.41) are $s = -\frac{1}{2}, -\frac{1}{2} - 1, \ldots$ of order 1 each; $s = -1 - \nu, -1 - \nu - 1, \ldots, 1$ or $s = \mu - 1, \mu - 2, \ldots 1$ are of order one each, and the poles $s = 0, -1, -2, \ldots$ are of order two each. Let us again denote the sum of the residues by S_j, $j = 1, 2, 3$. The sum of the residues corresponding to the poles $s = -\frac{1}{2}, -\frac{1}{2} - 1, \ldots$ remains the same as before and it is equal to the following:

$$S_1 = \Gamma\left(-\frac{1}{2}\right) \Gamma\left(\frac{1}{2}+\nu\right) \left(\frac{z^2}{4}\right)^{\frac{1}{2}} {}_0F_2\left(-; \frac{3}{2}, \frac{1}{2}-\nu; -\frac{z^2}{4}\right). \quad (2.76)$$

The other set of poles of order 1 each are $s = -1 - \nu - r, r = 0, 1, \ldots, -\nu - 2$ and the sum of the residues, denoted by S_2, is the following:

$$S_2 = \sum_{r=0}^{-\nu-2} \lim_{s \to -1-\nu-r} (s+1+\nu+r)\Gamma(s+\nu+1)\Gamma(s)\Gamma\left(\frac{1}{2}+s\right)\left(\frac{z^2}{4}\right)^{-s}$$

$$= \left(\frac{z^2}{4}\right)^{1+\nu} \sum_{r=0}^{-\nu-2} \left(\frac{z^2}{4}\right)^r \frac{(-1)^r}{r!} \Gamma(-\nu-1-r)\Gamma\left(-\frac{1}{2}-\nu-r\right)$$

$$= \left(\frac{z^2}{4}\right)^{1+\nu} \Gamma(-\nu-1)\Gamma\left(-\frac{1}{2}-\nu\right) \sum_{r=0}^{-\nu-2} \frac{(-1)^r}{r!} \left(\frac{z^2}{4}\right)^r \frac{1}{(\nu+2)_r (\nu+3/2)_r}.$$
$$(2.77)$$

Poles of order 2 are at $s = -r, r = 0, 1, 2, \ldots$. Consider the relation,

$$(s+r)^2 = \frac{(s+r)^2 (s+r-1)^2 \cdots (s)^2 (s-1) \cdots (s-\nu+1)}{(s+r-1)^2 \cdots s^2 (s-1) \cdots (s-\nu+1)}.$$

Then

$$(s+r)^2 \Gamma(s)\Gamma(s+\nu+1) = \frac{\Gamma^2(s+r+1)}{(s+r-1)^2 \cdots s^2(s-1) \cdots (s-\nu+1)}.$$

Let

$$B'_r = \lim_{s \to -r} B'(s) = \lim_{s \to -r} [(s+r)^2 \Gamma(s)\Gamma(s+\nu+1)\Gamma(1+s)]$$

$$= \frac{\Gamma^2(1)\Gamma\left(\frac{1}{2}-r\right)}{(-1)^2(-2)^2 \cdots (-r)^2(-r-1) \cdots (-r+\nu+1)}$$

$$= \frac{\Gamma^2(1)\Gamma\left(\frac{1}{2}-r\right)(-1)^{\nu+1}}{r!(r-\nu-1)!}. \qquad (2.78)$$

Let

$$A'_r = \lim_{s \to -r} A'(s) = \lim_{s \to -r} \left\{ \frac{\partial}{\partial s} \ln B'(s) \right\}$$

$$= \psi(r+1) + \psi(r-\nu) + \psi\left(\frac{1}{2}-r\right). \qquad (2.79)$$

Thus, the sum of the residues corresponding to the poles $s = -r$, $r = 0, 1, \ldots$, by using steps similar to the ones employed before, and denoted by S_3, is the following:

$$S_3 = \sum_{0}^{\infty} \left(\frac{z^2}{4}\right)^r \left\{-\ln\left(\frac{z^2}{4}\right) + A'_r\right\} B'_r, \qquad (2.80)$$

where B'_r and A'_r are defined in (2.78) and (2.79) respectively. With this, we have the following theorem:

Theorem 2.5 *For ν a negative integer,*

$$\pi^{-\frac{1}{2}}(S_1 + S_2 + S_3) = \pi^{-\frac{1}{2}} G_{0,3}^{3,0}\left[\frac{z^2}{4}\bigg|_{0,\frac{1}{2},1+\nu}\right]$$

$$= \pi^{-\frac{1}{2}}\left(\frac{z^2}{4}\right)^{1+\nu} \Gamma(-\nu-1)\Gamma\left(-\frac{1}{2}-\nu\right)$$

$$\times \sum_{r=0}^{-\nu-2} \frac{(-1)^r}{r!}\left(\frac{z^2}{4}\right)^r \frac{1}{(\nu+2)_r(\nu+3/2)_r}$$

$$- 2\Gamma\left(\frac{1}{2}+\nu\right)\left(\frac{z^2}{4}\right)^{\frac{1}{2}} {}_0F_2\left(-;\frac{3}{2},\frac{1}{2}-\nu;-\frac{z^2}{4}\right)$$

$$+ \pi^{-\frac{1}{2}} \sum_{r=0}^{\infty} \left(\frac{z^2}{4}\right)^r \left\{-\ln\left(\frac{z^2}{4}\right) + A'_r\right\} B'_r,$$

where B'_r and A'_r are defined in (2.78) and (2.79), respectively.

2.10 Series Representations for the Thermonuclear Functions: $G_{0,3}^{3,0}(\cdot)$

Case 2.11 ν a positive half integer

Let $\nu = m + \frac{1}{2}, m = 0, 1, \ldots$. Then

$$\Gamma(s)\Gamma(s+\nu+1)\Gamma\left(\frac{1}{2}+s\right) = \Gamma(s)\Gamma\left(s+m+\frac{3}{2}\right)\Gamma\left(s+\frac{1}{2}\right)$$

and the poles of the Meijer's G-function in (2.67) are the following: $s = -r, r = 0, 1, \ldots$ are of order 1 each; $s = -\frac{1}{2} - r, r = 0, 1, \ldots, m$ are of order 1 each; $s = -m - \frac{3}{2} - r, r = 0, 1, \ldots$ are of order 2 each. The sum of the residues corresponding to the poles $s = -r, r = 0, 1, \ldots$ is the following, denoted again by S_1:

$$S_1 = \sum_{r=0}^{\infty} \frac{(-1)^r}{r!} \left(\frac{z^2}{4}\right)^r \Gamma\left(\frac{1}{2}-r\right)\Gamma\left(m+\frac{3}{2}-r\right)$$

$$= \Gamma\left(\frac{1}{2}\right)\Gamma\left(m+\frac{3}{2}\right)\sum_{r=0}^{\infty} \frac{(-1)^r}{r!} \frac{1}{(1/2)_r(-m-1/2)_r}\left(\frac{z^2}{4}\right)^r$$

$$= \Gamma\left(\frac{1}{2}\right)\Gamma\left(m+\frac{3}{2}\right) {}_0F_2\left(-;\frac{1}{2},-m-\frac{1}{2};-\frac{z^2}{4}\right). \quad (2.81)$$

Evidently, the sum of the residues corresponding to the poles $s = -\frac{1}{2} - r, r = 0, 1, \ldots, m$, denoted by S_2, is the following:

$$S_2 = \sum_{r=0}^{m} \frac{(-1)^r}{r!} \left(\frac{z^2}{4}\right)^{\frac{1}{2}+r} \Gamma\left(-\frac{1}{2}-r\right)\Gamma(1+m-r)$$

$$= \Gamma\left(-\frac{1}{2}\right)\Gamma(1+m)\left(\frac{z^2}{4}\right)^{\frac{1}{2}} \sum_{r=0}^{m} \frac{(-1)^r}{r!}\left(\frac{z^2}{4}\right)^r \frac{1}{(3/2)_r(-m)_r}. \quad (2.82)$$

Let

$$C(s) = \left(s+m+\frac{3}{2}+r\right)^2 \Gamma\left(\frac{1}{2}+s\right)\Gamma\left(\frac{3}{2}+m+s\right)\Gamma(s).$$

As done in the previous cases, insert the following factors in the numerator and in the denominator.

$$\left(s+m+\frac{3}{2}+r-1\right)^2 \cdots \left(s+m+\frac{3}{2}\right)^2 \left(s+m+\frac{3}{2}-1\right) \cdots \left(s+\frac{1}{2}\right)$$

and write

$$C(s) = \frac{\Gamma^2\left(s+m+\frac{3}{2}+r+1\right)\Gamma(s)}{\left(s+m+\frac{3}{2}+r-1\right)^2 \cdots \left(s+m+\frac{3}{2}\right)^2 \left(s+m+\frac{3}{2}-1\right) \cdots \left(s+\frac{1}{2}\right)}.$$

Now take the limit and denote the result as C_r, that is,

$$C_r = \lim_{s \to -m-\frac{3}{2}-r} C(s)$$

$$= \frac{\Gamma^2(1)\Gamma\left(-m - \frac{3}{2} - r\right)}{(-1)^2 \cdots (-r)^2(-r-1) \cdots (-m-1-r)}$$

$$= \frac{(-1)^{m+1+r}\Gamma\left(-m - \frac{3}{2}\right)}{r!(m+1+r)!(m+5/2)_r}. \tag{2.83}$$

Let $D(s) = \frac{\partial}{\partial s} \ln C(s)$ and

$$D_r = \lim_{s \to -m-\frac{3}{2}-r} D(s) = \psi(m+1) + \psi(m+r+2) + \psi\left(-m - \frac{3}{2} - r\right). \tag{2.84}$$

Hence, the sum of the residues corresponding to the poles of order 2 each, denoted by S_3, is the following:

$$S_3 = \sum_{r=0}^{\infty} \left(\frac{z^2}{4}\right)^{m+\frac{3}{2}+r} \left\{-\ln\left(\frac{z^2}{4}\right) + D_r\right\} C_r. \tag{2.85}$$

Therefore, we have the following theorem:

Theorem 2.6 *For ν a positive half integer, namely $\nu = m + \frac{1}{2}$, $m = 0, 1, \ldots$*

$$\pi^{-\frac{1}{2}}(S_1 + S_2 + S_3) = \pi^{-\frac{1}{2}} G_{0,3}^{3,0}\left[\frac{z^2}{4}\bigg|_{0,\frac{1}{2},1+\nu}\right]$$

$$= \Gamma\left(m + \frac{3}{2}\right) {}_0F_2\left(-; \frac{1}{2}, -m - \frac{1}{2}; -\frac{z^2}{2}\right)$$

$$- 2\Gamma(1+m)\left(\frac{z^2}{4}\right)^{\frac{1}{2}} \sum_{r=0}^{m} \frac{(-1)^r}{r!}\left(\frac{z^2}{4}\right)^r \frac{1}{(3/2)_r(-m)_r}$$

$$\pi^{-\frac{1}{2}}\left(\frac{z^2}{4}\right)^{m+\frac{3}{2}} \sum_{r=0}^{\infty} \left(\frac{z^2}{4}\right)^r \left\{-\ln\left(\frac{z^2}{4}\right) + D_r\right\} C_r,$$

where C_r and D_4 are defined in (2.83) and (2.84), respectively.

Case 2.12 ν a negative half integer

Let $\nu = -m - \frac{1}{2}$, $m = 0, 1, 2, \ldots$. In this case,

$$\Gamma(s)\Gamma\left(s + \frac{1}{2}\right)\Gamma(s + \nu + 1) = \Gamma(s)\Gamma\left(s + \frac{1}{2}\right)\Gamma\left(s + \frac{1}{2} - m\right).$$

2.10 Series Representations for the Thermonuclear Functions: $G_{0,3}^{3,0}(\cdot)$

The poles at $s = -r, r = 0, 1, 2, \ldots$ are of order 1 each; the poles at $s = m - \frac{1}{2} - r, r = 0, 1, \ldots, m - 1$ are of order 1 each and the poles at $s = -\frac{1}{2} - r, r = 0, 1, 2, \ldots$ are of order 2 each. Again, we will denote the sum of the residues by $S_j, j = 1, 2, 3$. The sum of the residues corresponding to the poles $s = -r, r = 0, 1, \ldots$, denoted by S_1, is the following:

$$S_1 = \sum_{r=0}^{\infty} \frac{(-1)^r}{r!} \left(\frac{z^2}{4}\right)^r \Gamma\left(\frac{1}{2} - r\right) \Gamma\left(\frac{1}{2} - m - r\right)$$

$$= \Gamma\left(\frac{1}{2}\right) \Gamma\left(\frac{1}{2} - m\right) \sum_{r=0}^{\infty} \frac{(-1)^r}{r!} \left(\frac{z^2}{4}\right)^r \frac{1}{(1/2)_r (m+1/2)_r}$$

$$= \Gamma\left(\frac{1}{2}\right) \Gamma\left(\frac{1}{2} - m\right) {}_0F_2\left(-; \frac{1}{2}, m + \frac{1}{2}; -\frac{z^2}{4}\right). \tag{2.86}$$

For the sum of the residues at the poles $s = m - \frac{1}{2} - r, r = 0, 1, \ldots, m - 1$, we have the following, denoted by S_2:

$$S_2 = \sum_{r=0}^{m-1} \frac{(-1)^r}{r!} \left(\frac{z^2}{4}\right)^{r-m+\frac{1}{2}} \Gamma(m - r) \Gamma\left(m - \frac{1}{2} - r\right)$$

$$= \left(\frac{z^2}{4}\right)^{-m+\frac{1}{2}} \Gamma(m) \Gamma\left(m - \frac{1}{2}\right) \sum_{r=0}^{m-1} \frac{(-1)^r}{r!} \left(\frac{z^2}{4}\right)^r \frac{1}{(-m+1)_r (-m+3/2)_r}. \tag{2.87}$$

Let

$$C'(s) = \left(s + \frac{1}{2} + r\right)^2 \Gamma\left(s + \frac{1}{2}\right) \Gamma\left(s + \frac{1}{2} - m\right) \Gamma(s).$$

As done in the previous cases, insert the following factors in the numerator and in the denominator

$$\left(s + \frac{1}{2} + r - 1\right)^2 \cdots \left(s + \frac{1}{2}\right)^2 \left(s - \frac{1}{2}\right) \cdots \left(s + \frac{1}{2} - m\right)$$

and write

$$C's(s) = \frac{\Gamma^2\left(s + \frac{1}{2} + r + 1\right)}{\left(s + \frac{1}{2} + r - 1\right)^2 \cdots \left(s + \frac{1}{2}\right)^2 \left(s - \frac{1}{2}\right) \cdots \left(s + \frac{1}{2} - m\right)}.$$

Now, take the limit of the logarithmic derivative of $C'(s)$ and write

$$D'_r = \lim_{s \to -\frac{1}{2} - r} \frac{\partial}{\partial s} \ln C'(s) = \psi(1) + \psi(m + r + 1) + \psi\left(-\frac{1}{2} - r\right). \tag{2.88}$$

Let

$$C'_r = \lim_{s \to -\frac{1}{2}-r} C'(s) = \frac{\Gamma^2(1)\Gamma\left(-\frac{1}{2}-r\right)}{(-1)^2(-2)^2 \cdots (-r)^2(-r-1) \cdots (-r-m)}$$
$$= \frac{(-1)^{m+r}\Gamma\left(-\frac{1}{2}\right)}{r!(m+r)!(3/2)_r}. \tag{2.89}$$

Now, we can write the sum of the residues corresponding to the poles at $s = -\frac{1}{2} - r, r = 0, 1, \ldots$ as the following, denoted by S_3:

$$S_3 = \sum_{r+0}^{\infty} \left(\frac{z^2}{4}\right)^{r+\frac{1}{2}} \left\{-\ln\left(\frac{z^2}{4}\right) + D'_r\right\} C'_r, \tag{2.90}$$

where D'_r and C'_r are given in (2.89) and (2.90), respectively. Hence, we have the following theorem:

Theorem 2.7 *For ν a negative half integer, namely $\nu = -m - \frac{1}{2}, m = 0, 1, \ldots$*

$$\pi^{-\frac{1}{2}}(S_1 + S_2 + S_3) = \pi^{-\frac{1}{2}} G_{0,3}^{3,0}\left[\frac{z^2}{4}\bigg|_{0,\frac{1}{2},1+\nu}\right]$$
$$= \Gamma\left(\frac{1}{2} - m\right){}_0F_2\left(-;\frac{1}{2}, m+\frac{1}{2}; -\frac{z^2}{4}\right)$$
$$+ \pi^{-\frac{1}{2}}\left(\frac{z^2}{4}\right)^{-m+\frac{1}{2}} \Gamma(m)\Gamma\left(m - \frac{1}{2}\right) \sum_{r=0}^{m-1} \frac{(-1)^r}{r!}\left(\frac{z^2}{4}\right)^r$$
$$\times \frac{1}{(-m+1)_r(-m+3/2)_r}$$
$$+ \pi^{-\frac{1}{2}}\left(\frac{z^2}{4}\right)^{\frac{1}{2}} \sum_{r=0}^{\infty}\left(\frac{z^2}{4}\right)^r \left\{-\ln\left(\frac{z^2}{4}\right) + D'_r\right\} C'_r,$$

where D'_r and C'_r are given in (2.88) and (2.89), respectively.

2.11 Series Representations for the Thermonuclear Functions: $G_{1,3}^{3,0}(\cdot)$

In view of the integral of the nonresonant nuclear reaction rate with modified Maxwell-Boltzmann distribution in (2.32) we have to consider the special case $n = 1, m = 2$, and $a = 1$ in (2.33):

2.11 Series Representations for the Thermonuclear Functions: $G_{1,3}^{3,0}(\cdot)$

$$\pi^{-\frac{1}{2}}d^{-\rho+1}\sum_{r=0}^{\infty}\frac{(-d)^r}{r!}G_{1,3}^{3,0}\left[\frac{z^2}{4d}\bigg|_{-\rho+r+1,0,\frac{1}{2}}^{-\rho+r+2}\right]$$

$$=\pi^{-\frac{1}{2}}d^{-\rho+1}\sum_{r=0}^{\infty}\frac{(-d)^r}{r!}\frac{1}{2\pi i}\int_{t-i\infty}^{t+i\infty}ds\,\frac{\Gamma(-\rho+r+1+s)\Gamma(s)\Gamma\left(\frac{1}{2}+s\right)}{\Gamma(-\rho+r+2+s)}\left(\frac{z^2}{4d}\right)^{-s}$$

$$=\pi^{-\frac{1}{2}}d^{-\rho+1}\sum_{r=0}^{\infty}\frac{(-d)^r}{r!}\frac{1}{2\pi i}\int_{t-i\infty}^{t+i\infty}ds\,\frac{\Gamma(s)\Gamma\left(\frac{1}{2}+s\right)}{(-\rho+r+1+s)}\left(\frac{z^2}{4d}\right)^{-s}. \qquad (2.91)$$

For the Mellin-Barnes integral representation of Meijer's G-function see the details in Mathai and Saxena (1973).

Case 2.13 $-\rho+r+1\neq\frac{\lambda}{2},\lambda=0,1,2,\ldots$

Note that the poles of the integrand in (2.91) are at $s=\rho-r-1$, $s=0,-1,-2,\ldots$ and $s=-\frac{1}{2},-\frac{1}{2}-1,\ldots$. Hence if $-\rho+r+1\neq\frac{\lambda}{2},\lambda=0,1,2,\ldots$ all the poles are simple and the following are the residues:

at $s=\rho-r-1$, is $\Gamma(\rho-r-1)\Gamma\left(\rho-r-\frac{1}{2}\right)\left(\frac{z^2}{4d}\right)^{-\rho+r+1}$

at $s=-\nu$, is $\dfrac{(-1)^\nu \Gamma\left(\frac{1}{2}-r\right)}{\nu!(-\rho+r+1-\nu)}\left(\frac{z^2}{4d}\right)^{\nu}$

at $s=-\frac{1}{2}-\nu$, is $\dfrac{(-1)^\nu \Gamma\left(-\frac{1}{2}-\nu\right)}{\nu!\left(-\rho+r+\frac{1}{2}-\nu\right)}\left(\frac{z^2}{4d}\right)^{\frac{1}{2}+\nu}$.

Hence, we have the following result:

Theorem 2.8 *For* $z>0, d>0, -\rho+r+1\neq\frac{\lambda}{2},\lambda=0,1,2,\ldots$

$$\pi^{-\frac{1}{2}}d^{-\rho+1}\sum_{r=0}^{\infty}\frac{(-d)^r}{r!}G_{1,3}^{3,0}\left[\frac{z^2}{4d}\bigg|_{-\rho+r+1,0,\frac{1}{2}}^{-\rho+r+2}\right]$$

$$=\pi^{-\frac{1}{2}}d^{-\rho+1}\sum_{r=0}^{\infty}\frac{(-d)^r}{r!}\bigg\{\Gamma(\rho-r-1)\Gamma(\rho-r-1/2)\left(\frac{z^2}{4d}\right)^{-\rho+r+1}$$

$$+\sum_{\nu=0}^{\infty}\frac{(-1)^\nu\Gamma\left(\frac{1}{2}-\nu\right)}{\nu!(-\rho+r+1-\nu)}\left(\frac{z^2}{4}\right)^{\nu}+\sum_{\nu=0}^{\infty}\frac{(-1)^\nu\Gamma\left(-\frac{1}{2}-\nu\right)}{\nu!(-\rho+r-\nu+1/2)}\left(\frac{z^2}{4d}\right)^{\frac{1}{2}+\nu}\bigg\}.$$

Case 2.14 $-\rho+r-1=\mu, \mu=0,1,2,\ldots$

If $-\rho+r-1=\mu, \mu=0,1,2,\ldots$ then at $s=-\mu$ there is a pole of order two in the integrand of (2.91). The remaining poles are simple. The residue at $s=-\mu, \mu=0,1,2,\ldots$, denoted by S_1, is therefore the following:

$$S_1 = \lim_{s \to -\mu} \frac{\partial}{\partial s} \left\{ \frac{(s+\mu)^2 C(s)}{(s+\mu)} \right\}, C(s) = \Gamma(s)\Gamma\left(\frac{1}{2}+s\right)\left(\frac{z^2}{4d}\right)^{-s},$$

$$= \left(\frac{z^2}{4d}\right)^{\mu} \left\{ -\ln\left(\frac{z^2}{4d}\right) + A \right\} B, \text{ where}$$

$$B = \lim_{s \to -\mu} \left\{ \frac{\Gamma(s+\mu+1)\Gamma(s+1/2)}{(s+\mu-1)\cdots(s)} \right\} = \frac{(-1)^{\mu}}{\mu!}\Gamma\left(\frac{1}{2}-\mu\right), \quad (2.92)$$

and

$$A = \lim_{s \to -\mu} \frac{\partial}{\partial s} \ln\left\{ \frac{\Gamma(s+\mu+1)\Gamma(s+1/2)}{(s+\mu-1)\cdots(s)} \right\}$$

$$= \psi(\mu+1) + \psi\left(\frac{1}{2}-\mu\right) \quad (2.93)$$

where $\psi(z)$ is the psi function $\psi(z) = \frac{\partial}{\partial s}\ln\Gamma(z)$ (Mathai and Saxena 1973). Thus, we can give the following result:

Theorem 2.9 *For* $z > 0, d > 0, -\rho + r + 1 = \mu, \mu = 0, 1, 2, \ldots$

$$\pi^{-\frac{1}{2}}d^{-\rho+1} \sum_{r=0}^{\infty} \frac{(-d)^r}{r!} G_{1,3}^{3,0}\left[\frac{z^2}{4d}\Big|_{-\rho+r+1,0,\frac{1}{2}}^{-\rho+r+2}\right]$$

$$= \pi^{-\frac{1}{2}}d^{-\rho+1} \sum_{r=0}^{\infty} \frac{(-d)^r}{r!} \Big\{ \sum_{\nu=0,\neq\mu}^{\infty} \frac{(-1)^\nu \Gamma(-\nu+1/2)}{\nu!(-\rho+r+1-\nu)} \left(\frac{z^2}{4d}\right)^\nu$$

$$+ \sum_{\nu=0}^{\infty} \frac{(-1)^\nu \Gamma(-\nu-1/2)}{\nu!(-\rho+r-\nu+1/2)} \left(\frac{z^2}{4d}\right)^{\frac{1}{2}+\nu} + \left(\frac{z^2}{4d}\right)^\mu \left\{-\ln\left(\frac{z^2}{4d}\right) + A\right\} B \Big\},$$
$$(2.94)$$

where A and B are given in (2.93) and (2.92), respectively.

Case 2.15 $-\rho + r - 1 = \frac{\lambda}{2}, \lambda = 1, 3, 5, \ldots$

If $-\rho + r - 1 = \frac{\lambda}{2}, \lambda = 1, 3, 5, \ldots$ then there is a pole of order two at $s = -\frac{\lambda}{2}$. The residue at $s = -\frac{\lambda}{2}$ reduces to the following, denoted by S_1:

$$S_1 = \lim_{s \to -\lambda/2} \frac{\partial}{\partial s} \left\{ \frac{(s+\lambda/2)^2 \Gamma(s)\Gamma(s+1/2)}{(s+\lambda/2)} \left(\frac{z^2}{4d}\right)^{-s} \right\}$$

$$= \left(\frac{z^2}{4d}\right)^{\lambda/2} \left\{ -\ln\left(\frac{z^2}{4d}\right) + A_1 \right\} B_1$$

where

$$B_1 = \frac{(-1)^{\lambda_1}}{\lambda_1!}\Gamma\left(-\frac{\lambda}{2}\right), \quad \frac{\lambda}{2} = \lambda_1 + \frac{1}{2}, \quad (2.95)$$

and
$$A_1 = \psi(\lambda_1 + 1) + \psi\left(-\frac{\lambda}{2}\right). \tag{2.96}$$

Theorem 2.10 *For* $z > 0, d > 0, -\rho + r + 1 = \frac{\lambda}{2}, \lambda = 1, 3, 5, \ldots, \frac{\lambda}{2} = \lambda_1 + \frac{1}{2}$

$$\pi^{-\frac{1}{2}} d^{-\rho+1} \sum_{r=0}^{\infty} \frac{(-d)^r}{r!} G_{1,3}^{3,0}\left[\frac{z^2}{4d} \Big|_{-\rho+r+1,0,\frac{1}{2}}^{-\rho+r+2}\right]$$
$$= \pi^{-\frac{1}{2}} d^{-\rho+1} \sum_{r=0}^{\infty} \frac{(-d)^r}{r!} \Big\{ \sum_{\nu=0, \neq \lambda_1}^{\infty} \frac{(-1)^\nu \Gamma(-\nu - 1/2)}{\nu!(-\rho - \nu + r + 1/2)} \left(\frac{z^2}{4d}\right)^{\frac{1}{2}+\nu}$$
$$+ \sum_{\nu=0}^{\infty} \frac{(-1)^\nu \Gamma(-\nu + 1/2)}{\nu!(-\rho + r - \nu + 1)} \left(\frac{z^2}{4d}\right)^\nu + \left(\frac{z^2}{4d}\right)^{\frac{\lambda}{2}} \left\{-\ln\left(\frac{z^2}{4d}\right) + A_1\right\} B_1\Big\},$$

where B_1 *and* A_1 *are given in (2.95) and (2.96), respectively.*

Note that in all the Theorems 2.8, 2.9, and 2.10, the inner series are nothing but a hypergeometric series of the type $_1F_2(z)$ and hence convergent for all z. Then, the outer sum in these theorems is dominated by a $_1F_1(z)$ which is also convergent for all values of the variable z. Hence, the series forms on the right side in Theorems 2.8, 2.9, and 2.10 are convergent (cp. Haubold and Mathai 1986a).

References

Atkinson, R.D'E., Houtermans, F.G.: Z. Phys. **54**, 656–665 (1929)
Bahcalll, J.N.: Astrophys. J. **143**, 259–261 (1966)
Bethe, H.A.: Les Prix Nobel 1967. Almqvist and Wiksells, Stockholm 1967 (cp. also Die Naturwissenschftern) **55**, 405–413 (1968)
Biswas, S., Ramadurai, S., Vahla, M.N.S.: Nucleosynthesis. Tata Institute of Fundamental Research Bombay, Bombay 1981. In: Proceedings of the Summer Workshop held in TIFR, Bombay during May 19–23 (1980)
Blatt, J.M., Weisskopf, V.F.: Theoretische Kernphysik. B.G. Teubner Verlangsgesellschaft, Leipzig (1959)
Burbidge, E.M., Burbidge, G.R., Fowler, W.A., Hoyle, F.: Rev. Mod. Phys. **29**, 547–650 (1957)
Critchfield, C.L.: Cosmology, Fusion and Other Matters. George Gamow Memorial Volume. In: Reines, F. (ed.) University of Colorado Press, Colorado, pp. 186–191 (1972)
Fowler, W.A.: Q.J.R.A.S. **15**, 82–106 (1974); The George Darwin Lecture (1973)
Fowler, W.A.: Rev. Mod. Phys. **56**, 149–179 (1984); The Nobel Lecture (1983)
Fowler, W.A., Caughlan, G.R., Zimmerman, B.A.: Ann. Rev. Astron. Astrophys. **5**, 525–570 (1967)
Fowler, W.A., Caughlan, G.R., Zimmerman, B.A.: Ann. Rev. Astron. Astrophys. **13**, 69–112 (1975)
Gamow, G., Teller, E.: Phys. Rev. **53**, 608–609 (1938)
Haubold, H.J., John, R.W.: Astron. Nachr. **300**, 63–75 (1979); **303**, 217 (1982)
Haubold, H.J., John, R.W.: Astron. Nachr. **303**, 161–187 (1982)
Haubold, H.J., Mathai, A.M.: Ann. Phys. (Leipzig) **41**, 380–396 (1984)
Haubold, H.J., Mathai, A.M.: Studies. Appl. Math. **75**, 123–137 (1986a)

Haubold, H.J., Mathai, A.M.: J. Math. Phys. **27**, 2203–2207 (1986b)
Haubold, H.J., Mathai, A.M.: J. Appl. Math. Phys. (ZAMP) **37**, 685–695 (1986c)
Mahai, A.M.: Math. Nachr. **48**, 129–139 (1971)
Mathai, A.M., Saxena, R.K.: Generalized Hypergeometric Functions with Applications in Statistics and Physical Sciences. Lecture Notes in Mathematics, vol. 348. Springer, Berlin-Heidelberg, New York (1973)
Salpeter, E.E.: Phys. Rev. **88**, 547–553 (1952)
Saxena, R.K.: Proc. Nat. Acad. Sci. India **26**, 400–413 (1960)
Wagoner, R.V.: Ap. J. Suppl. **18**, 247–295 (1969)

Open Access This chapter is licensed under the terms of the Creative Commons Attribution 4.0 International License (http://creativecommons.org/licenses/by/4.0/), which permits use, sharing, adaptation, distribution and reproduction in any medium or format, as long as you give appropriate credit to the original author(s) and the source, provide a link to the Creative Commons license and indicate if changes were made.

The images or other third party material in this chapter are included in the chapter's Creative Commons license, unless indicated otherwise in a credit line to the material. If material is not included in the chapter's Creative Commons license and your intended use is not permitted by statutory regulation or exceeds the permitted use, you will need to obtain permission directly from the copyright holder.

Chapter 3
The Solar Neutrino Problem

3.1 Introduction: The Solar Neutrino Problem

"...some say the solar neutrino problem is still with us. Others say there never was a problem. My purpose is to present this ambiguous situation to you in such a way that you can make your own judgement." (Fowler 1977). It is not the aim of the present chapter to add one more suggestion to the long list of attempts for the solution of the so-called solar neutrino problem (for an updated list cf. Haxton 1984).

It is the merit of R. DAVIS, Jr. and his associates to have developed a radiochemical method of detecting the solar neutrino via the reaction $\nu_e + {}^{37}Cl \rightarrow {}^{37}Ar + e^-$ over a period of about 30 years (Bahcall and Davis 1982). This detection method is sensitive primarily to the high energy neutrinos from the decay ${}^8B \rightarrow {}^8Be^* + e^+ + \nu_e$, where ${}^8B$ is produced in the proton-proton chain with the overall net result: $4p + e^- \rightarrow \alpha + e^+ + 2\nu_e + 26.73 MeV$. The branching ratios of the proton-proton chain are very sensitive to the solar temperature and require a detailed model of the evolution and internal structure of the Sun (Haubold and Gerth 1983). The observations yield, as an average of the runs from 1970 to 1983, a rate $\sum_i (\phi_i \sigma_i)_{obs} = (2.1 \pm 0.3) SNU (1\ SNU = 10^{-36}$ captures s^{-1} per ${}^{37}Cl$ atom (Davis et al. 1984). The rate predicted by the standard solar model is $\sum_i (\phi_i \sigma_i)_{theor} = (7.6 \pm 3.3) SNU (3\sigma)$ (Bahcall et al. 1982). The disagreement between standard theory and observations by a factor of four constitutes what has come to be known as the solar neutrino problem in the sense questioned by FOWLER as mentioned above. Over more than 15 years a large number of more or less fundamental solutions, modifying either the neutrino physics, nuclear physics or the solar model construction, have been proposed. The present chapter is an attempt to construct a simple analytic model to study the internal structure and the neutrino emission of the Sun in its present stage of evolution.

In Sect. 3.2, the nuclear reactions of the proton-proton chain and its branching are considered and the total nuclear energy generation rate of the proton-proton chain under the assumption of near-statistical equilibrium between the reactions will be derived. In Sect. 3.3, an analytic model for the central nuclear burning region of the

Sun is constructed by taking into account the assumptions of mass conservation, hydrostatic equilibrium, and energy conservation. These calculations assume a non-linear density distribution for the solar model and employ the equation of state of the perfect gas. What concerns the method of solar model construction is that in some sense we are going back to the method of Chandrasekhar (1939) and Hayashi et al. (1962). In Sect. 3.4, the basic integral of the solar nuclear energy generation and solar neutrino emission is derived taking into account the closed-form representation of the nonresonant thermonuclear reaction rate (Haubold and Mathai 1984). That integral is the analytic equivalent to the rough estimation of the solar neutrino flux by $N_\nu \approx 2L_\odot/E$, where $L_\odot$ is the luminosity of the Sun and E is the energy liberated by the overall net reaction as mentioned above, not including the energy of the two neutrinos ($N_\nu \approx 2 \times 10^{38} s^{-1}$). In Sect. 3.5, the integral of the nuclear energy generation and solar neutrino emission is evaluated analytically in closed-form by means of modern results of the integration theory of generalized hypergeometric functions (cp. Mathai and Saxena 1973, 1978). In bringing together the results of the analytic evaluations described in Sects. 3.3, 3.4, 3.5, respectively, analytic results for the solar neutrino emission rates are represented in Sect. 3.10.

3.2 The Proton-Proton Chain

The calculation of the rate of energy generation and neutrino production from the proton-proton chain is not so simple as in the case of the CNO cycles, for example, because of the three possible modes of termination of the chain (cf. Table 3.1, column (i) and (ii), respectively). The locally liberated heat, Q, varies from $Q_{ppIII} = 19.1 MeV$ to $Q_{ppI} = 26.2 MeV$ because of different neutrino energy losses. The individual exoergic reaction energies shown in Table 3.1 (column (i)) include positron annihilation but do not include neutrino energy. The rate of energy generation is not simply proportional to the rate of the first reaction (1) in the proton-proton chain, the $^1H(p, e^+\nu_e)\,^2H$ reaction, but depends also on density, temperature and the abundances of the various nuclei that enter into the various reactions. The temperature of the deep interior of the Sun is high enough, that all reactions in each mode of termination can occur in time scales short compared to the age of the Sun. The only exception is the proton-proton reaction itself (cf. Table 3.1). Thus, one can assume that a near-statistical equilibrium among the intermediate products of the modes will be established, in which the reaction rates of creation and destruction of each nucleus are approximately equal. Under those conditions, the calculation of energy generation and neutrino production for the full chain as shown in Table 3.1 is much simplified (Fowler 1977; Kavanagh 1982; Bahcall et al. 1982). The branching of the various reactions of the proton-proton chain is shown in columns (ii) and (iii), respectively, in Table 3.1. For the principal mode (PPI) terminating with reaction (3) we have the rate of energy generation per unit mass,

3.2 The Proton-Proton Chain

Table 3.1 Basic facts for the proton-proton chain (adapted from KAVANAGH (1982), BAHCALL et al. (1982), and HAXTON (1984))

(i)	(ii)	(**)(iii)	(**)(iv)	(**)(v)	(**)(vi)	(**)(vii)
Reactions, except for reaction energies not including positron annihilation but not including neutrino energy	Termination and locally contained heat	Branching percentages	$S_{11}(0)$[keVb]	$\tau_D(0)$[years]	$E_{\text{cut-max}}$[keV]	$E_{\nu,\text{max}}$ [MeV]
(0) $^1\text{H} + e^- + {}^1\text{H} \rightarrow {}^2\text{H} + \nu_e$	PPI	0.25%				1.44 (mono)
(1) $^1\text{H} + {}^1\text{H} \rightarrow {}^2\text{H} + e^+ + \nu_e + 1.19$ MeV		99.75%	$S_{11}(0) = 3.88 \times 10^{-22}$	5.8×10^9		0.42
(2) $^2\text{H} + {}^1\text{H} \rightarrow {}^3\text{He} + \gamma + 5.49$ MeV	$Q_{\text{PPI}} = 26.2$ MeV		$S_{12}(0) 2.5 \times 10^{-4}$	3.2×10^{-15}	15	
(3) $^3\text{He} + {}^3\text{He} \rightarrow {}^4\text{He} + 2{}^1\text{H} + 12.85$ MeV	40%		$S_{33}(0) 4.7 \times 10^3$	1.5×10^5	33	
(4) $^3\text{He} + {}^4\text{He} \rightarrow {}^7\text{Be} + \gamma + 1.54$ MeV		31%	$S_{34}(0) 0.52$	6.5×10^7	107	
(5) $^7\text{Be} + e^- \rightarrow {}^7\text{Li} + \nu_e + \gamma + 0.05$ MeV	PPII	99.7%		0.2		0.86 (90%) (mono)
(6) $^7\text{Li} + {}^1\text{H} \rightarrow {}^4\text{He} + {}^4\text{He} + 17.34$ MeV	$Q_{\text{PPII}} = 25.7$ MeV			2×10^{-4}		0.38 (10%)
(7) $^7\text{Be} + {}^1\text{H} \rightarrow {}^8\text{B} + \gamma + 0.14$ MeV	PPIII	0.3%	$S_{17}(0) 2.9 \times 10^{-2}$	71	117	
(8) $^8\text{B} \rightarrow {}^8\text{Be} + e^+ + \nu_e + 7.7$ MeV	$Q_{\text{PPIII}} = 19.1$ MeV			3×10^{-8}		14.06
(9) $^8\text{Be} \rightarrow {}^4\text{He} + {}^4\text{He} + 3.0$ MeV				10^{-19}		

$$S_{PPI} = \frac{1}{\rho} Q_{PPI} \frac{1}{2} n_3 n_3 <\sigma v>_{33}, \qquad (3.1)$$

where n_3 is the number density of ^3He nuclei and $<\sigma v>_{33}$ is the thermally averaged product of the cross section σ for the reaction and relative velocity v of the interacting particles. The mode PPI is accompanied by neutrino production in the reactions (1), $^1H(p, e^+ \nu_e) \, ^2H$, and (0), $^1H(e^- p, \nu_e) \, ^2H$, respectively. In the mode PPII under the assumption of near-statistical equilibrium, the net rate of creation of ^4He nuclei is equal to the rate of any of the reactions occurring in the mode PPII (the same holds for PPIII). Choosing the first reactions of each of the modes PPII and PPIII, we get for the rates of energy generation per unit mass

$$\epsilon_{PPII} = \frac{1}{\rho} Q_{PPII} \, n_7 p_{7e^-}, \qquad (3.2)$$

and

$$\epsilon_{PPIII} = \frac{1}{\rho} Q_{PPIII} \, n_7 n_1 <\sigma v>_{71}, \qquad (3.3)$$

where n_7 and n_1 are the number densities of ^7Be and ^1H, respectively, and p_{7e^-} is the probability per unit time per nucleus of electron capture, $<\sigma v>_{71}$ is the thermally averaged product of the cross section σ for the reaction (7) and relative velocity v of the interacting particles. The neutrino producing reactions of PPII and PPIII are the reactions (5) and (8), respectively.

For PPI the condition of near-statistical equilibrium means that

$$\frac{1}{2} n_1 n_1 <\sigma v>_{11} - 2\left(\frac{1}{2}\right) n_3 n_3 <\sigma v>_{33} - n_3 n_4 <\sigma v>_{34} = 0, \qquad (3.4)$$

where n_4 is the number density of ^4He nuclei, the factors $(1/2)$ account for the fact that a reaction between identical particles always involves two of these particles. The factor 2 in (3.57) accounts for the fact that two ^3He nuclei are consumed in each reaction (3). If the reaction $^3He(\alpha, \gamma) \, ^7Be$ would be negligible one simply states $\frac{1}{2} n_3 n_3 <\sigma v>_{33} = \left(\frac{1}{2}\right) \frac{1}{2} n_1 n_1 <\sigma v>_{11}$ which is correct if only the PPI mode operates. Similarly, for the reactions following the formation of ^7Be we have

$$n_3 n_4 <\sigma v>_{34} - n_7 p_{7e^-} - n_7 n_1 <\sigma v>_{71} = 0. \qquad (3.5)$$

For the total rate of energy generation per unit mass from all three modes PPI, PPII, PPIII we find

$$\epsilon_{PP} = \epsilon_{PPI} + \epsilon_{PPII} + \epsilon_{PPIII}, \qquad (3.6)$$

which we can write with (3.1), (3.2), (3.3), and by the help of (3.5),

3.2 The Proton-Proton Chain

$$\epsilon_{PP} = \frac{1}{\rho}\Big[Q_{PPI}\frac{1}{2}n_3n_3 <\sigma v>_{33} + Q_{PPII}n_3n_4 <\sigma v>_{34}$$
$$+ (Q_{PPIII} - Q_{PPII})n_7n_1 <\sigma v>_{71}\Big]. \tag{3.7}$$

Equation (3.7) is the total nuclear energy generation rate for the proton-proton chain expressed in terms of thermonuclear reaction rates of the reactions (3), (4), and (7) shown in Table 3.1. Taking into account (3.57) we can also write ϵ_{PP} in (3.7) in terms of the reaction (1), $^1H(p,e^+\nu_e)\,^2H$, which determines the overall thermonuclear energy generation of the proton-proton chain, the reaction (4), $^3He(\alpha,\gamma)\,^7Be$, which is the branching reaction between the modes PPI and PPII/PPIII in the proton-proton chain, and the reaction (7), $^7Be(p,\gamma)\,^8B$, which is the first reaction of the mode PPIII containing the high energy neutrino producing reaction of the decay $^8B \to\,^8B_e^* + e^+ + \nu_e$. As we know, the neutrinos from the decay of ^8B dominate the capture rate in the solar neutrino experiment of DAVIS and associates (Fowler 1977; Bahcall et al. 1982; Bahcall and Davis 1982).

What concerns the neutrino producing reactions in the proton-proton chain is that the beta decay reactions (1) and (8) in Table 3.1 produce neutrinos in an allowed FERMI distribution with maximum energy $E_{\nu_{max}}$ as shown in column (vii) of Table 3.1. However, the flux of ^8B neutrinos deviates from an allowed FERMI distribution because the ^8B_e final state populated in this decay is a broad resonance where $E_{\nu_{max}}$ is computed for the center of this resonance (Fowler 1977; BAHCALL et al. 1982). The electron capture reactions (0) and (5) produce line sources of neutrinos of maximum energy $E_{\nu_{max}}$ also shown in Table 3.1 column (vii). The electron capture of ^7Be leads partially to the ground state of ^7Li (90%) and partially to an excited state (10%) of ^7Li (Fowler 1977; Bahcall et al. 1982).

One of the most serious problems in the theory of solar nuclear energy generation and solar neutrino production is the dependence of both on the cross sections of the nuclear reactions involved. It is quite interesting that the first reaction of the proton-proton chain, $^1H(p,e^+\nu_e)\,^2H$ has never been directly observed in the nuclear laboratory because of its extremely small cross section. This reaction can occur only if the two protons are brought together by a nuclear collision. During the extremely short time of encounter between the protons one or both must have a chance to beta decay to become a neutron, a positron, and an electron neutrino. The neutron can then be captured by the second proton to form a deuteron which is a very rare event. May be this reaction will never be observed in the laboratory. However, the present state of the theory of beta decay makes it possible to estimate the cross section and thus the cross section factor with high precision. Practically, the same holds for the reaction (0) shown in Table 3.1 (ep. Bahcall et al. 1982). For all other nuclear reactions shown in Table 3.1, the nuclear cross section can be measured in the nuclear laboratory, but, at least to large energies in comparison to the order of the central temperature of the Sun ($T_c \approx 15.5 \times 10^6 K$, $kT_c \approx 1.34$ keV; cf. Table 3.1). It is generally accepted that the necessary extrapolation of the experimental cross sections measured for energies up till $E_{lab_{max}}$ as shown in column (vi) of Table 3.1 to solar energies is well understood theoretically (Fowler 1977; Kavanagh 1982; Bahcall et al. 1982; Haxton

1984). If one removes the strong energy dependence of the nuclear cross section due to the COULOMB barrier and displays the experimental data in the form of the astrophysical nuclear cross section factor $S(E)$, where E is the center-of-mass energy, one obtains the extrapolated S-factors at zero energy as shown in column (iv) of Table 3.1 (Bahcall et al. 1982).

3.3 An Analytic Model for the Central Region of the Sun

The equations of the internal structure of the Sun—mass conservations, hydrostatic equilibrium, energy conservation, and energy transport - form a system of nonlinear differential equations, three relations which characterize specifically the behaviour of the interior of the Sun, and particular conditions for the boundaries. A complete solution of that boundary value problem can only be obtained by numerical integration techniques. However, the aim of the present Section is to construct an analytic solar model by separating the condition of hydrostatic equilibrium from the consideration of the energy transport inside the solar material (cp. Haubold and Mathai 1984; Haubold and Mathai 1986). As it is well-known from the computation of detailed standard solar models, the temperature of the Sun increases towards the center, but does not attain the value necessary for thermonuclear energy generation until about $r = 0.2R_\odot (M = 0.35M_\odot, L = 0.95L_\odot$; cf. Table 3.2; Bahcall et al. 1982). Thus, throughout the region from $r = R_\odot$ to $r = 0.2R_\odot$, we can adopt a constant value of the luminosity L. For the treatment of the energy generation by nuclear reactions in

Table 3.2 Basic parameters for the solar model (adopted from BAHCALL et al. (1982))

Solar parameter(*)		Value
Luminosity	$L_\odot$	3.86×10^{33} erg s^{-1}
Mass	$M_\odot$	1.99×10^{33} g
Radius	$R_\odot$	6.96×10^{10} cm
Central hydrogen abundance by mass	X_c	0.355
Central helium abundance by mass	Y_c	0.6222
Central heavy elements abundances by mass	Z_c	0.0228
mean molecular weight for central solar conditions	μ_c	0.8417
Standard solar model parameters(*)		
Central density	$\rho_{c\odot}$	156 gcm^{-3}
Central pressure	$P_{c\odot}$	2.39×10^{17} dyncm^{-2}
Central temperature	$T_{c\odot}$	15.5×10^6 K
Analytic solar model parameters		
Central density	$\rho_{c\odot}$	3.52 gcm^{-3}
Central pressure	$P_{c\odot}$	1.31×10^{17} dyncm^{-2}
Central temperature	$T_{c\odot}$	9.7×10^6 K

3.3 An Analytic Model for the Central Region of the Sun

the deep interior of the Sun, we take into consideration the hydrostatic equilibrium and the energy conservation which determine the changes in the solar central conditions. We regard the density distribution $\rho(r)$ of our stellar model as an arbitrary but specified function of the distance parameter r,

$$\rho(r) = \rho_0 \left[1 - \left(\frac{r}{R_\odot} \right)^\delta \right], \delta > 0 \tag{3.8}$$

which is capable to reproduce the solar density distribution in the central region by choosing the free parameter δ; $\rho(r = 0) = \rho_c$ denotes the central stellar matter density, $\rho(R_\odot) = 0$. For reasons of symmetry of the solar model we take $\delta = 2$. From the equation of mass conservation,

$$\frac{d}{dr} M(r) = 4\pi r^2 \rho(r), \tag{3.9}$$

we obtain the mass distribution with (3.8):

$$M(r) = 4\pi \rho_c \int_0^r dt\, t^2 \left[1 - \left(\frac{t}{R_\odot} \right)^\delta \right] = \frac{4\pi}{3} \rho_c r^3 \left[1 - \frac{3}{(\delta + 3)} \left(\frac{r}{R_\odot} \right)^\delta \right]. \tag{3.10}$$

Note that from (3.10) one can get the central density, that is,

$$\rho(r = 0) = \rho_c = \frac{3}{4\pi} \frac{(\delta + 3)}{\delta} \frac{M_\odot}{R_\odot^3}.$$

Putting $\delta = 2$ we have the solar case for (7.3),

$$M(r) = \frac{5}{2} M_\odot \left(\frac{r}{R_\odot} \right)^3 \left[1 - \frac{3}{5} \left(\frac{r}{R_\odot} \right)^2 \right], \rho_{c\odot} = \frac{15}{8\pi} \frac{M_\odot}{R_\odot^3}, \tag{3.11}$$

the respective central density $\rho_{c\odot}$ for the analytic solar model is given in Table 3.2.

The equation of hydrostatic equilibrium between the total pressure per unit volume and the gravity per unit volume

$$\frac{d}{dr} P(r) = -\frac{GM(r)\rho(r)}{r^2}, \tag{3.12}$$

has the following solution with (3.8) and (3.10):

$$P(r) = P(0) - \int_0^r dt \frac{GM(t)\rho(t)}{t^2}$$

$$= \frac{4\pi G}{3}\rho_c^2 R_\odot^2 \left\{ \xi - \frac{1}{2}\left(\frac{r}{R_\odot}\right)^2 + \frac{(\delta+6)}{(\delta+2)(\delta+3)}\left(\frac{r}{R_\odot}\right)^{\delta+2} \right.$$

$$\left. - \frac{3}{2(\delta+1)(\delta+3)}\left(\frac{r}{R_\odot}\right)^{2\delta+2} \right\}, \qquad (3.13)$$

where we took into account the boundary conditions,

$$P(r = R_\odot) = 0; \; P(r=0) = P_c = \frac{3G}{4\pi}\xi\frac{(\delta+3)^2}{\delta^2}\frac{M_\odot^2}{R_\odot^4};$$

G is the gravitational constant, and

$$\xi = \frac{1}{2} - \frac{(\delta+6)}{(\delta+2)(\delta+3)} + \frac{3}{2(\delta+1)(\delta+3)}. \qquad (3.14)$$

For our analytic solar model we obtain with $\delta = 2$ for (3.13):

$$P(r) = 5P_c\left\{\frac{1}{5} - \frac{1}{2}\left(\frac{r}{R_\odot}\right)^2 + \frac{2}{5}\left(\frac{r}{R_\odot}\right)^4 - \frac{1}{10}\left(\frac{r}{R_\odot}\right)^6\right\},$$

$$P_{c\odot} = \frac{15}{16\pi}G\frac{M_\odot^2}{R_\odot^4}, \qquad (3.15)$$

the respective value for the central pressure is given in Table 3.2.

Thus, (3.9), (3.12), and the boundary conditions mentioned above determine the pressure distribution (3.13), corresponding to the given density distribution (3.8), for which hydrostatic equilibrium will be obtained. In the range of temperatures of the interior of the Sun and because we restrict ourselves to a star of solar mass we can make the restriction that the radiation pressure inside the solar material is negligible (Chandrasekhar 1939). According to the perfect gas law, in the simplest form suggested by the kinetic theory of gases, we have the pressure

$$P(r) = \frac{kN_A}{\mu}\rho(r)T(r). \qquad (3.16)$$

In (3.16), μ is the mean molecular weight, N_A is AVOGADRO's constant, and k is the BOLTZMANN constant. With (3.16) the temperature distribution $T(r)$ corresponding to hydrostatic equilibrium is determined by (3.8) and (3.13):

$$T(r) = \frac{\mu}{kN_A}\frac{P(r)}{\rho(r)} = \frac{4\pi G}{3kN_A}\mu\rho_c R_\odot^2 \left[1 - \left(\frac{r}{R_\odot}\right)^\delta\right]^{-1} \{\xi - \frac{1}{2}\left(\frac{r}{R_\odot}\right)^2$$
$$+ \frac{(\delta+6)}{(\delta+2)(\delta+3)}\left(\frac{r}{R_\odot}\right)^{\delta+2} - \frac{3}{2(\delta+1)(\delta+3)}\left(\frac{r}{R_\odot}\right)^{2\delta+2}\}, \quad (3.17)$$

and the central temperature is given by

$$T(r=0) = T_c = \frac{G}{kN_A}\xi\frac{(\delta+3)}{\delta}\mu\frac{M_\odot}{R_\odot},$$

where ξ is given by (3.14).

In the case $\delta = 2$ of our analytic solar model we have for (3.17),

$$T(r) = 5T_c\left[1 - \left(\frac{r}{R_\odot}\right)^2\right]^{-1}\{\frac{1}{5} - \frac{1}{2}\left(\frac{r}{R_\odot}\right)^2 + \frac{2}{5}\left(\frac{r}{R_\odot}\right)^4 - \frac{1}{10}\left(\frac{r}{R_\odot}\right)^6\},$$
$$T_{c\odot} = \frac{1}{2}\frac{G}{kN_A}\mu\frac{M_\odot}{R_\odot}, \quad (3.18)$$

and the corresponding central temperature $T_{c\odot}$ is given in Table 3.2.

3.4 Solar Thermonuclear Energy Generation: Energy Conservation and Solar Luminosity

The analytic solar model constructed in Sect. 3.7 specified the distribution of mass and pressure for a given density distribution and required the temperature distribution to give hydrostatic equilibrium. This temperature distribution will in general not lead to thermal equilibrium. However, aside from the question of energy transport the solar model will be in complete thermal equilibrium only if the equation of energy conservation is satisfied:

$$\frac{d}{dr}L(r) = 4\pi r^2 \rho(r)\epsilon(r). \quad (3.19)$$

Equation (3.19) states that the net increase in the rate of energy flow from the inside to the outside of a spherical shell of the Sun is equal to the rate of energy production within the shell. In (3.19), $L(r)$ represents the energy flux through the sphere with radius r, $\epsilon(r)$ is the rate of thermonuclear energy generation per unit mass and includes the tiny energy losses via solar neutrinos. The net outflow of energy per second, $L(r)$, through the sphere of radius r is determined, in the case of radiative transfer in solar matter, by the local values of the opacity and the temperature gradient. However, with our assumed density distribution (3.8) the energy transport equation can be satisfied at only one point of the Sun. This serious assumption is justified by the aim of the present chapter to investigate the physical conditions in the central nuclear burning region of the Sun where neutrinos are generated. Note that the luminosity,

$$L(R_\odot) = \int_0^{R_\odot} dr\, 4\pi r^2 \rho(r) \epsilon(r), \tag{3.20}$$

is not completely insensitive to the energy generation rate but, basically determined by the mass of the star (Chandrasekhar 1939). The quantity $L(r)$ remains constant and will be equal to its value at the surface of the Sun, $L(R_\odot)$, as long as one remains outside the central region where nuclear energy generation takes place. If we are concerned with only one specific reaction $1 + 2 \to 3 + 4$, then we have the "internal luminosity" going back to the specific nuclear reaction in question:

$$L_{12}(R_\odot) = \int_0^{R_\odot} dr\, 4\pi r^2 \rho(r) \epsilon_{12}(r). \tag{3.21}$$

The energy generation rate per unit mass $\epsilon_{12}(r)$ is, beside the equation of state (3.16) (the opacity is not specified in our solar model by assumption), one of the three material equations of our solar model. Let E_{12} denote the amount of energy given off in a single reaction of the proton-proton chain written in the standard notation $1 + 2 \to 3 + 4$, where 1 and 2 denote the incoming particles, and 3 and 4 denote the outgoing particles.

In the more or less developed analytic theory of the internal structure of the Sun one refers in general to the definition of the thermonuclear energy generation rate in the following form:

$$\epsilon(r) = \epsilon_0(\rho_0, T_0) \left(\frac{\rho(r)}{\rho_0}\right)^\alpha \left(\frac{T(r)}{T_0}\right)^\beta, \quad \alpha, \beta \text{ real}. \tag{3.22}$$

In (8.4) the thermonuclear energy generation rate is expressed in terms of powers of the density and temperature, where the subscript 0 designates central conditions, α and β are constants, and ϵ_0 contains the chemical composition (Chandrasekhar 1939; Hayashi, Hōshi, and Sugimoto 1962; Haubold and Mathai 1984, 1986). In general, the representation (3.22) of $\epsilon(r)$ does not describe the energy generation rate of a specific reaction, but, the total nuclear energy generation rate of a chain of nuclear reactions like the proton-proton chain.

In the following we do not refer to the representation (3.22), but, take into consideration the definition of the nuclear energy generation rate,

$$\epsilon_{12}(r) = \frac{1}{\rho(r)} E_{12}\, r_{12}(\rho(r), T(r)), \tag{3.23}$$

containing the thermonuclear reaction rate $r_{12}(\rho(r), T(r))$ whose theory can be formulated on the basis of physical principles (cf. e.g., Parker et al. 1964; Haubold and Mathai 1984; Haubold and John 1981).

3.5 The Thermonuclear Reaction Rate

All reactions involved in the proton-proton chain are nonresonant. In the calculations of the energy generation and neutrino emission via the proton-proton chain described in Sect. 3.2 we will adopt the following definition of the nonresonant nuclear cross section (Salpeter 1952; Parker et al. 1964; Haubold and Mathai 1984):

$$\sigma(E) = \frac{S(E)}{E} \exp\{-2\pi \eta(E)\}, \tag{3.24}$$

where $\eta(E)$ is the SOMMERFELD parameter, given by,

$$\eta(E) = \left(\frac{\mu^*}{2}\right)^{\frac{1}{2}} \frac{Z_1 Z_2 e^2}{h E^{1/2}}, \tag{3.25}$$

where Z_1 and Z_2 are the charges of the interacting particles, E is the center-of-mass energy, e is the quantum of electric charge, h is PLANCK's quantum of action, $\mu^* = m_1 m_2/(m_1 + m_2) = A_1 A_2/(A_1 + A_2) N_A$ is the reduced mass, A_1 and A_2 are the atomic mass numbers of the particles. As mentioned in Sect. 3.2, the cross section for the nuclear reactions that occur in the proton-proton chain cannot generally be measured at the energies of interest for solar conditions (cp. Table 3.1 column (vi)). Hence for the cross section factor $S(E)$ in (3.24), extrapolations from higher energy measurements must be used to obtain a zero energy intercept, $S(0)$, and average values for the first and second derivatives at low energies, $S'(0)$ and $S''(0)$. Therefore, $S(E)$ is expanded by using a MACLAURIN series in the light of the weak dependence of $S(E)$ on the relative kinetic energy of the particles (Bahcall 1966; Critchfield 1972; Haubold and Mathai 1984)

$$S(E) = S(0) + S'(0)E + S''(0)E^2. \tag{3.26}$$

All the relevant cross section factors for the proton-proton chain evaluated at zero energy, $S(0)$, are shown in Table 1 column (iv) (adopted from Bahcall et al. 1982).

The MAXWELL-BOLTZMANN averaged thermonuclear reaction rate of a nonresonant nuclear reaction r_{12} in (3.23) can now be written (HAUBOLD and MATHAI 1984):

$$r_{12} = \left(1 - \frac{1}{2}\delta_{12}\right) n_1 n_2 \left(\frac{8}{\pi \mu^*}\right)^{\frac{1}{2}} \frac{1}{(kT)^{3/2}} \sum_{\nu=0}^{2} \frac{S^\nu(0)}{\nu!}$$
$$\times \int_0^\infty dE\, E^\nu \exp\left\{-\left(\frac{E}{kT} + 2\pi \eta(E)\right)\right\}, \tag{3.27}$$

where δ_{12} is the KRONECKER symbol, n_1 and n_2 are the particle number densities. The number densities are more explicitly written as

$$n_i(r) = N_A \frac{X_i}{A_i} \rho(r), \tag{3.28}$$

where X_i is the atomic abundance by mass of the nuclei of type i. In connection with the explicit form of the thermonuclear reaction rate in (3.27) it is convenient to introduce the quantity $<\sigma v>_{12}$, defined by the relation

$$r_{12} = \left(1 - \frac{1}{2}\delta_{12}\right) n_1 n_2 <\sigma v>_{12}, \tag{3.29}$$

where $<\sigma v>_{12}$ is the thermally averaged product of the cross section σ for the reaction and relative velocity v of the interacting particles. The mean lifetime, $r_2(1)$, of nucleus 1 for interaction with nucleus 2 is given as follows:

$$\lambda_2(1) = \frac{1}{\tau_2(1)} = n_2 <\sigma v>_{12} = \rho N_A \frac{X_2}{A_2} <\sigma v>_{12}, \tag{3.30}$$

where $\lambda_2(1)$ is the decay rate of 1 for interaction with 2.

The mean lifetime, $\tau_2(1)$ defined in (3.30), for the reactions of the proton-proton chain are given in Table 3.1 column (v) for central physical conditions of the standard model of the Sun (cp. Kavanagh 1982).

After defining the basic operations (3.23), (3.29) and (3.30) for the description of the dynamics of nuclear reactions in the deep interior of the Sun we find the final representation of the thermonuclear reaction rate by substituting (3.28) in (3.27) and writing $y = E/(kT)$:

$$r_{12}(r) = \left(1 - \frac{1}{2}\delta_{12}\right) N_A^{\frac{5}{2}} \frac{X_1 X_2 (A_1 + A_2)^{\frac{1}{2}} 8^{\frac{1}{2}}}{(A_1 A_2)^{3/2}} \frac{1}{\pi} \rho^2(r)$$

$$\times \sum_{\nu=0}^{2} (kT(r))^{\nu-\frac{1}{2}} \frac{S^\nu(0)}{\nu!} N_{i_\nu}(z(r)), \tag{3.31}$$

where

$$N_{i_\nu}(z(r)) = \int_0^\infty dy \, y^\nu e^{-y-z(r)y^{-\frac{1}{2}}} = \frac{1}{\pi^{\frac{1}{2}}} G_{0,3}^{3,0} \left[\frac{z^2(r)}{4} \Big|_{1+\nu,\frac{1}{2},0} \right], \tag{3.32}$$

$$z(r) = 2\pi \left(\frac{\mu^*}{2kT(r)}\right)^{\frac{1}{2}} \frac{Z_1 Z_2 e^2}{h}. \tag{3.33}$$

Equation (3.32) is the basic closed-form representation of the nonresonant thermonuclear reaction rate integral by means of MEIJER's G-function (cf. Mathai and Saxana 1973; Haubold and Mathai 1984). This closed-form representation is appropriate to perform analytical operations and from (3.32) approximate expressions easily follow for small and large values of the characteristic parameter $z(r)$ in (3.33),

3.6 The Neutrino Emission Rate

that is the COULOMB barrier energy divided by thermal energy, which appears in the argument of the G-function (Haubold and John 1981; Haubold and Mathai 1984).

For large values of $z(r)$ we get the asymptotic representation for the G-function in (3.32) (cf. Mathai and Saxena 1973)

$$G_{0,3}^{3,0}\left[\frac{z^2(r)}{4}\bigg|_{1+\nu,\frac{1}{2},0}\right] \to 2\left(\frac{\pi}{3}\right)^{\frac{1}{2}}\left(\frac{z(r)}{2}\right)^{\frac{2\nu+1}{3}}\exp\left\{-3\left(\frac{z(r)}{2}\right)^{\frac{2}{3}}\right\}. \quad (3.34)$$

This relation reproduces results from the well-known papers of Salpeter (1952, for $\nu = 0$) and Bahcall (1966, for $\nu = 0, 1, 2$). Inserting (8.16) for $N_{i_\nu}(z(r))$ in (3.31) we get

$$r_{12}(r) = \left(1 - \frac{1}{2}\delta_{12}\right) N_A^{\frac{5}{2}} \frac{X_1 X_2 (A_1 + A_2)^{\frac{1}{2}}}{(A_1 A_2)^{3/2}} \frac{2^{\frac{5}{2}}}{3^{1/2}\pi} \rho^2(r)$$

$$= \exp\left\{-3\left(\frac{z(r)}{2}\right)^{\frac{2}{3}}\right\} \sum_{\nu=0}^{2} (kT(r))^{\nu-\frac{1}{2}} \frac{S^{(\nu)}(0)}{\nu!} \left(\frac{z(r)}{2}\right)^{\frac{2\nu+1}{2}}. \quad (3.35)$$

Once again, it is obvious that if $S(E)$ in (3.26) is nearly constant most of the value of (3.35) comes from values of E near zero. The first approximation for well-behaved cross section factors is to treat $S(E)$ as a constant, $S(0)$, defined as the value at $E = 0$. Hence, for $\nu = 0$ we obtain

$$r_{12}(r) = \left(1 - \frac{1}{2}\delta_{12}\right) N_A^{\frac{5}{2}} \frac{X_1 X_2 (A_1 + A_2)^{\frac{1}{2}}}{(A_1 A_2)^{3/2}} \frac{2^{\frac{5}{2}}}{3^{1/2}\pi} \rho^2(r)$$

$$\times \frac{S(0)}{(kT(r))^{1/2}} \left(\frac{z(r)}{2}\right)^{\frac{1}{2}} \exp\left\{-3\left(\frac{z(r)}{2}\right)^{\frac{2}{3}}\right\}. \quad (3.36)$$

that reproduces the asymptotic representation of the thermonuclear reaction rate underlying the fundamental papers of Fowler (1984; cf. also Parker et al. 1964; Critchfield 1972; Haubold and Mathai 1984).

3.6 The Neutrino Emission Rate

In the following evaluations we do not take into account the asymptotic form (3.36) of the thermonuclear reaction rate but the closed-form representation (3.31) with (3.32) for $\nu = 0$, and write (3.21) in terms of $r_{12}(\rho(r), T(r))$,

$$L_{12}(R_\odot) = \int_0^{R_\odot} dr \, 4\pi r^2 E_{12} r_{12}(\rho(r), T(r)). \quad (3.37)$$

If we divide the "internal luminosity" $L_{12}(R_\odot)$ by the amount of energy E_{12}, then we get the total number of particles per second N_{12} liberated in the reaction $1 + 2 \to 3 + 4$ in question, that is,

$$\frac{L_{12}(R_\odot)}{E_{12}} = N_{12} = 4\pi \int_0^{R_\odot} dr\, r^2\, r_{12}(\rho(r), T(r))$$

$$= 4\pi \left(1 - \frac{1}{2}\delta_{12}\right) N_A^{\frac{5}{2}} \frac{X_1 X_2 (A_1 A_2)^{\frac{1}{2}}}{(A_1 + A_2)^{3/2}} \frac{8^{\frac{1}{2}}}{\pi} S_{12} R_\odot^3$$

$$\times \int_0^1 dx\, x^2 \rho^2(R_\odot x)(kT(R_\odot x))^{-\frac{1}{2}} \int_0^\infty dy\, e^{-y - zy^{-\frac{1}{2}}(kT(R_\odot x))^{-\frac{1}{2}}},$$
(3.38)

where $x = r/R_\odot$ and

$$z = 2\pi \left(\frac{\mu^*}{2}\right)^{\frac{1}{2}} \frac{Z_1 Z_2 e^2}{h} = E_G^{\frac{1}{2}},$$

where E_G is the GAMOW energy. The flux of solar neutrinos at the Earth due to the reaction $1 + 2 \to 3 + 4$ in the Sun can then be written

$$\phi_{12} = \frac{N_{12}}{4\pi(AU)^2} = \frac{L_{12}(R_\odot)}{4\pi(AU)^2 E_{12}}, \quad (3.39)$$

where AU abbreviates the mean distance between the Earth and the Sun (Astronomical Unit $= 1.496 \times 10^{13}$ cm) and N_{12} is also called the neutrino emission rate.

3.7 The Integral for the Solar Nuclear Energy Generation and the Solar Neutrino Fluxes: The General Case of the Basic Integral

Consider the following integral,

$$Q_\delta = \int_0^R dr\, r^p [\rho(r)]^q [kT(r)]^{-t} \int_0^\infty dy\, e^{-ay - z[kT(r)]^{-b} y^{-\frac{n}{m}}}, \quad (3.40)$$

for the case (3.8) of the density distribution and (3.17) of the temperature distribution, respectively; $\delta > 0, p > 0, t > 0, a > 0, q > 0, z > 0$. For convenience the inner integral in (3.40) will be rewritten in terms of the MELLIN-BARNES type integral by using a result in Haubold and Mathai (1984) which will be stated here as a lemma (cf. also Haubold and Mathai 1987).

3.7 The Integral for the Solar Nuclear Energy Generation and the Solar Neutrino Fluxes ... 65

Lemma 3.1 *For* $\alpha > 0, \beta > 0, u > 0, \Re(1 - \gamma + s) > 0$,

$$\int_0^\infty dt\, t^{-\gamma} e^{-\alpha t - u t^{-\beta}} = \frac{\alpha^{\gamma-1}}{\beta} \frac{1}{2\pi i} \int_{c-i\infty}^{c+i\infty} ds\, \Gamma(s/\beta)\Gamma(1-\gamma+s) \left(\alpha u^{\frac{1}{\beta}}\right)^{-s}, \tag{3.41}$$

where $i^2 = -1$, $c > \Re(\gamma - 1)$, *and* $\Re(\cdot)$ *denotes the real part of* $(\cdot)$.

When $\beta = n/m$, $n, m = 1, 2, \ldots$ one has $\Gamma(s/\beta) = \Gamma(ms/n)$. If s/n is replaced by s' and if $\Gamma(ms')$ and $\Gamma(1 - \gamma + ns')$ are expanded by using the multiplication formula for gamma functions and by using the resulting quantity if the right hand side of (3.41) is written as a MEIJER's G-function then the result in (3.41) agrees with the result established by Saxena (1960) with the help of transform calculus. A proof of (3.41) by using statistical techniques is given in Haubold and Mathai (1984) and hence the proof won't be repeated here.

From (3.41) we may note that the right hand side of (3.41) is a H-function, see for example Mathai and Saxena (1978), and that it exists for all values of $\alpha u^{1/\beta} > 0$ and the inner integral in (3.40) for the case (3.8) is dominated by a beta type integral. Hence, by substituting (3.41) in (3.40) and by rewriting (3.40) one has the following:

$$Q_\delta = a^{-1}(m/n) \frac{1}{2\pi i} \int_{c-i\infty}^{c+i\infty} ds\, \Gamma(ms/n)\Gamma(1+s) \left(\alpha z^{1/\beta}\right)^{-s}$$

$$\times \int_0^\infty dr\, r^p [\rho(r)]^q [kT(r)]^{\frac{bs}{\beta} - t}, \tag{3.42}$$

where $\beta = n/m$.

Now the inner integral in (3.42) for the density distribution (3.8) and the temperature distribution (3.17) reduces to the following, denoting the integral by I_1:

$$I_1 = \int_0^R dr\, r^p [\rho(r)]^q [kT(r)]^{\frac{bs}{\beta} - t} = R^{p+1} \int_0^1 dx\, x^p [\rho(Rx)]^q [kT(Rx)]^{\frac{bs}{\beta} - t}$$

$$= R^{p+1} \rho_c^q \left[\frac{4\pi \mu G}{3 N_A k} \rho_c R^2\right]^{\frac{bs}{\beta} - t} \int_0^1 dx\, x^p (1 - x^\delta)^{q + t - \frac{bs}{\beta}}$$

$$\times \left\{\xi - \frac{1}{2}x^2 + \frac{(\delta + 6)}{(\delta + 2)(\delta + 3)} x^{\delta+2} - \frac{3}{2(\delta + 1)(\delta + 3)} x^{2\delta+2}\right\}^{\frac{bs}{\beta} - t}, \tag{3.43}$$

and where ξ is given by (3.14). Making the transformation $y = x^\delta$ and taking out ξ one has the following:

$$I_1 = \frac{R^{p+1} \rho_c^q}{\delta} \left[\frac{4\pi \mu G}{3 N_A k} \xi \rho_c R^2\right]^{\frac{bs}{\beta} - t} \int_0^1 dy\, y^{\frac{p+1}{\delta} - 1} (1-y)^{q+t-\frac{bs}{\beta}} [1 - u(y)]^{\frac{bs}{\beta} - t}, \tag{3.44}$$

where

$$u(y) = \xi^{-1} y^{\frac{2}{\delta}} \left\{ \frac{1}{2} - \frac{(\delta+6)}{(\delta+2)(\delta+3)} y + \frac{3}{2(\delta+1)(\delta+3)} y^2 \right\}, \quad (3.45)$$

where ξ is given in (3.14).

3.8 The Solar Case of the Basic Integral

In the case $\delta = 2$, $[1 - u(y)]$ of (3.44) reduces to the following form:

$$1 - u(y) = 1 - \xi^{-1} y \left(\frac{1}{2} - \frac{2}{5} y + \frac{1}{10} y^2 \right), \quad \xi = \frac{1}{5},$$

$$= 1 - \frac{5}{2} y + 2y^2 - \frac{1}{2} y^3$$

$$= (1-y)^2 \left(1 - \frac{1}{2} y \right).$$

Hence (3.44) reduces to the following form, denoted by I_2:

$$I_2 = \int_0^{R_\odot} dr \, r^p [\rho(r)]^q [kT(r)]^{\frac{bs}{\beta}-t} = \frac{1}{2} \rho_c^q R_\odot^{p+1} \left[\frac{4\pi \mu G}{15 N_A k} \rho_c R_\odot^2 \right]^{\frac{bs}{\beta}-t}$$

$$\times \int_0^1 dy \, y^{\frac{p+1}{2}-1} (1-y)^{q+\frac{bs}{\beta}-t} \left(1 - \frac{1}{2} y \right)^{\frac{bs}{\beta}-t}. \quad (3.46)$$

But

$$\left(1 - \frac{1}{2} y \right)^{-(t-\frac{bs}{\beta})} = \sum_{m_1=0}^{\infty} \frac{(t - bs/\beta)_{m_1}}{m_1!} \left(\frac{1}{2} \right)^{m_1} y^{m_1}.$$

Hence

$$\int_0^1 dy \, y^{\frac{p+1}{2}-1} (1-y)^{q+\frac{bs}{\beta}-t} \left(1 - \frac{1}{2} y \right)^{\frac{bs}{\beta}-t}$$

$$= \sum_{m_1=0}^{\infty} \frac{(t - bs/\beta)_{m_1}}{m_1!} \left(\frac{1}{2} \right)^{m_1} \frac{\Gamma(m_1 + (p+1)/2) \Gamma(1+q-t+bs/\beta)}{\Gamma(1+q-t+m_1+(p+1)/2+bs/\beta)}$$

by using a beta integral. Thus, we have for $\delta = 2$, noting that

$$\left(t - \frac{bs}{\beta} \right)_{m_1} = \frac{\Gamma(t + m_1 - bs/\beta)}{\Gamma(t - bs/\beta)},$$

3.8 The Solar Case of the Basic Integral

$$I_2 = \frac{1}{2}\rho_c^q R_\odot^{p+1} \left[\frac{4\pi\mu G}{15N_A k}\rho_c R_\odot^2\right]^{\frac{bs}{\beta}-t} \Gamma((p+1)/2) \sum_{m_1=0}^{\infty} \frac{((p+1)/2)_{m_1}}{m_1!}\left(\frac{1}{2}\right)^{m_1}$$

$$\times \frac{\Gamma(t+m_1-bs/\beta)\Gamma(q-t+1+bs/\beta)}{\Gamma(t-bs/\beta)\Gamma(1+q-t+m_1+(p+1)/2+bs/\beta)}. \tag{3.47}$$

Substituting (3.47) in (3.42) we have for $\delta = 2$, $\beta = n/m$,

$$Q_2 = a^{-1}\left(\frac{m}{n}\right)\frac{1}{2}R_\odot^{p+1}\rho_c^q\left[\frac{4\pi\mu G}{15N_A k}\rho_c R_\odot^2\right]^{-t}$$

$$\times \Gamma\left(\frac{p+1}{2}\right)\sum_{m_1=0}^{\infty}\frac{[(p+1)/2]_{m_1}(1/2)_{m_1}}{m_1!}$$

$$\times \frac{1}{2\pi i}\int_{L_1}ds\frac{\Gamma(ms/n)\Gamma(1+s)\Gamma(1+q-t+bsm/n)\Gamma(t+m_1-bsm/n)}{\Gamma(t-bsm/n)\Gamma(1+q-t+m_1+(p+1)/2+bsm/n)}d^{-s} \tag{3.48}$$

where L_1 is a suitable contour, and

$$d = az^{\frac{m}{n}}\left(\frac{15N_A k}{4\pi\mu\rho_c R_\odot^2 G}\right)^{\frac{mb}{n}}. \tag{3.49}$$

Note that the integral in (3.48) is nothing but a H-function, see Mathai and Saxena (1978). Hence, we have for $\delta = 2$,

$$Q_2 = a^{-1}\left(\frac{m}{n}\right)\frac{1}{2}R_\odot^{p+1}\rho_c^q\left[\frac{4\pi\mu G}{15N_A k}\rho_c R_\odot^2\right]^{-t}$$

$$\times \Gamma\left(\frac{p+1}{2}\right)\sum_{m_1=0}^{\infty}\frac{\left(\frac{p+1}{2}\right)_{m_1}\left(\frac{1}{2}\right)^{m_1}}{m_1!}$$

$$\times H_{2,4}^{3,1}\left[d\left|\begin{matrix}(1-t-m_1,1),(1+q-t+m_1+(p+1)/2,mb/n)\\(0,m/n),(1,1),(q-t+1,bm/n),(1-t,mb/n)\end{matrix}\right.\right]. \tag{3.50}$$

It is easy to note that the H-function exists for all values of $d > 0$ and that it behaves like the following, see Mathai and Saxena (1973). For small values of d it behaves like, $|d|^\alpha$, $\alpha = \min\left\{0, 1, \frac{n}{bm}(q-t+1)\right\} = 0$ for $q \geq t$. For large values of d it behaves like $|d|^\beta$ where $\beta = \max\left\{\frac{n}{mb}(-t-m_1)\right\} = -\frac{nt}{mb}$. Hence, using this fact we have the following result, noting that

$$\sum_{m_1=0}^{\infty}\frac{[(p+1)/2]_{m_1}(1/2)_{m_1}}{m_1!} = \left(1-\frac{1}{2}\right)^{-\frac{p+1}{2}} = 2^{\frac{p+1}{2}}.$$

3.9 Fox's H-Function

Definition

$$H_{p,q}^{m,n}(z) = H_{p,q}^{m,n}\left[z\Big|_{(b_q,B_q)}^{(a_p,A_p)}\right] = H_{p,q}^{m,n}\left[z\Big|_{(b_1,B_1),\ldots,(b_q,B_q)}^{(a_1,A_1),\ldots,(a_p,A_p)}\right] = \frac{1}{2\pi i}\int_L ds\, z^s \psi(s),$$

$$\psi(s) = \frac{\{\prod_{j=1}^m \Gamma(b_j - B_j s)\}\{\prod_{j=1}^n \Gamma(1 - a_j + A_j s)\}}{\{\prod_{j=m+1}^q \Gamma(1 - b_j + B_j s)\}\{\prod_{j=n+1}^p \Gamma(a_j - A_j s)\}}$$

where $0 \leq n \leq p$, $1 \leq m \leq q$; A_j, $j = 1, \ldots, p$ and B_j, $j = 1, \ldots, q$ are real positive numbers, a_j, $j = 1, \ldots, p$ and b_j, $j = 1, \ldots, q$ are complex numbers such that $A_j(b_h + \nu) \neq B_h(b_j - \lambda - 1)$, for $\nu, \lambda = 0, 1, 2, \ldots$; $h = 1, \ldots, m$; $j = 1, \ldots, n$. L is a contour separating the points $s = (b_j + \nu)/B_j$, $j = 1, \ldots, m$; $\nu = 0, 1, \ldots$ which are the poles of $\Gamma(b_j - B_j s)$, $j = 1, \ldots, m$, from the points $s = (a_j - \nu - 1)/A_j$, $j = 1, \ldots, n$; $\nu = 0, 1, \ldots$, which are the poles of $\Gamma(1 - a_j + A_j s)$, $j = 1, \ldots, n$. The H-function is an analytic function of z and makes sense if the following existence conditions are satisfied: (i) for all $z \neq 0$ with $\mu > 0$, and (ii) for $0 < |z| < \beta^{-1}$ with $\mu > 0$;

$$\mu = \sum_{j=1}^q B_j - \sum_{j=1}^p A_j, \quad \beta = \{\prod_{j=1}^p A_l^{A_j}\}\{\prod_{j=1}^q B_j^{-B_j}\}.$$

Asymptotic expansions: Orders for small and large values:

$$H_{p,q}^{m,n}(z) \to O(|z|^c), \text{ for small values of } z$$

for $\mu \geq 0$ and $c = \min\{\Re(b_j/B_j), j = 1, \ldots, m\}$; and

$$H_{p,q}^{m,n}(z) \to O(|z|^d) \text{ for large values of } z$$

for $\mu \geq 0$, $|\arg z| < \alpha\pi/2$, and $d = \max\{\Re((a_j - 1)/A_j), j = 1, \ldots, n\}$, where

$$\alpha = \sum_{j=1}^n A_j - \sum_{j=n+1}^p A_j + \sum_{j=1}^m B_j - \sum_{j=m+1}^q B_j;$$

$$\beta = \{\prod_{j=1}^p A_j^{A_j}\}\{B_j^{-B_j}\};$$

$$\gamma = \sum_{j=1}^q b_j - \sum_{j=1}^p a_j + \frac{p}{2} - \frac{q}{2};$$

$$\lambda = \sum_{j=1}^m B_j - \sum_{j=m+1}^q B_j - \sum_{j=1}^p A_j;$$

$$\mu = \sum_{j=1}^{q} B_j - \sum_{j=1}^{p} A_j.$$

3.10 Meijer's G-Function

This G-function is a special case of the H-function. When $A_j = 1, j = 1, \ldots, p$ and $B_j = 1, j = 1, \ldots, q$, one has

$$H_{p,q}^{m,n}\left[z \Big|_{(b_j,1), j=1,\ldots,q}^{(a_j,1), j=1,\ldots,p}\right] = G_{p,q}^{m,n}\left[z \Big|_{b_j, j=1,\ldots,q}^{a_j, j=1,\ldots,p}\right].$$

The definition as well as the asymptotic expansions for the G-function can be derived from those of the H-function.

Theorem 3.1 *For $\delta = 2$, Q_2 is approximated to the following:*

$$Q_2 \approx a^{-1}\left(\frac{m}{n}\right)\frac{1}{2}R_\odot^{p+1}\rho_c^q\left[\frac{4\pi\mu G}{15N_A k}\rho_c R_\odot^2\right]^{-t}\Gamma\left(\frac{p+1}{2}\right)2^{\frac{p+1}{2}}$$

for small values of d, and

$$Q_2 \approx a^{-1}\left(\frac{m}{n}\right)\frac{1}{2}R_\odot^{p+1}\rho_c^q\left[\frac{4\pi\mu G}{15N_A k}\rho_c R_\odot^2\right]^{-t}\Gamma\left(\frac{p+1}{2}\right)2^{\frac{p+1}{2}}d^{-\frac{nt}{mb}}$$

for large values of d, where d is defined in (3.49) (cf. also Haubold and Mathai 1987).

3.11 Analytic Results Connecting Solar Structure Parameters and Solar Neutrino Emission Rates

Neglecting the nuclear energy generation of the Sun via the CNO cycle, because it provides a negligible contribution to the total energy output of the Sun, we conclude from Eqs. (3.6), (3.7) and (3.20) that

$$\begin{aligned}L(R_\odot) &= 4\pi\int_0^{R_\odot} dr\, r^2\rho(r)(\epsilon_{PPI} + \epsilon_{PPII} + \epsilon_{PPIII})\\ &= 4\pi\frac{Q_{PPI}}{2}\int_0^{R_\odot} dr\, r^2 n_3 n_3 <\sigma v>_{33} + 4\pi Q_{PPII}\int_0^{R_\odot} dr\, r^2 n_3 n_4 <\sigma v>_{34}\\ &\quad + 4\pi(Q_{PPIII} - Q_{PPII})\int_0^{R_\odot} dr\, r^2 n_7 n_1 <\sigma v>_{71},\end{aligned} \quad (3.51)$$

or, we may write, with Eq. (3.1)

$$L(R_\odot) = 4\pi \frac{Q_{PPI}}{4} \int_0^{R_\odot} dr\, r^2 n_1 n_1 <\sigma v>_{11}$$
$$+ 4\pi \left(Q_{PPII} - \frac{1}{2} Q_{PPI}\right) \int_0^{R_\odot} dr\, r^2 n_3 n_4 <\sigma v>_{34}$$
$$+ 4\pi (Q_{PPIII} - Q_{PPII}) \int_0^{R_\odot} dr\, r^2 n_7 n_1 <\sigma v>_{71}. \qquad (3.52)$$

According to (3.38) and (3.6) also we obtain the neutrino fluxes of the nuclear reactions of the proton-proton chain in terms of the same quantities $<\sigma v>$ as included in (3.52).

$$\phi_{11} = \phi(pp) = \frac{N_{11}}{4\pi(AU)^2} = \frac{1}{2(AU)^2} \int_0^{R_\odot} dr\, r^2 n_1 n_1 <\sigma v>_{11}, \qquad (3.53)$$

$$\phi_{71} = \phi(^7Be) = \frac{N_{34} - N_{71}}{4\pi(AU)^2}$$
$$= \frac{1}{(AU)^2} \left\{ \int_0^{R_\odot} dr\, r^2 n_3 n_4 <\sigma v>_{34} - \int_0^{R_\odot} dr\, r^2 n_7 n_1 <\sigma v>_{71} \right\}, \qquad (3.54)$$

$$\phi_8 = \phi(^8B) = \frac{N_{71}}{4\pi(AU)^2} = \frac{1}{(AU)^2} \int_0^{R_\odot} dr\, r^2 n_7 n_1 <\sigma v>_{71}. \qquad (3.55)$$

With the representations of the nuclear output of the Sun by (3.52) and the respective neutrino fluxes by (3.53), (3.54), and (3.55) we traced back the computation of these quantities to the evaluation of the basic integral in (3.38). For the solar model discussed in Sect. 3.3 and the closed form representation of the nonresonant thermonuclear reaction rate obtained in Sect. 3.4 we get the neutrino emission rate in (3.38) and (3.50) for $a = 1, b = \frac{1}{2}, n = 1, m = 2, p = 2, q = 2$, and $t = \frac{1}{2}$.

$$N_{12} = 4\pi \left(1 - \frac{1}{2}\delta_{12}\right) N_A^{\frac{5}{2}} \frac{X_1 X_2 (A_1 + A_2)^{\frac{1}{2}}}{(A_1 A_2)^{3/2}} \frac{8^{\frac{1}{2}}}{\pi} S_{12}(0) Q_2, \qquad (3.56)$$

where

$$Q_2 = \left(\frac{\pi}{4}\right)^{\frac{1}{2}} \frac{R_\odot^3 \rho_{c\odot}^2}{T_{c\odot}^{1/2}} \sum_{m_1=0}^{\infty} \frac{(3.2)_{m_1} (1/2)^{m_1}}{m_1!} H_{2,4}^{3,1} \left[\frac{z^2}{T_c} \Big|_{(0,2),(1,1),(5/2,1),(1/2,1)}^{(1/2-m_1,1),(4+m_1,1)}\right] \qquad (3.57)$$

where z is given in (3.9). The value of he central matter density of the solar model is given in (3.8) and that for the central temperature in (3.18). According to the Theorem 3.1, the approximated value for Q_2 in (3.57) can now be written as

$$Q_2 \approx (2\pi)^{\frac{1}{2}} \frac{R_\odot^3 \rho_{c\odot}^2}{T_c^{1/2}} \qquad (3.58)$$

3.11 Analytic Results Connecting Solar Structure Parameters and ...

for small values of $d = z^2/T_c$ given in (9.10), and

$$Q_2 \approx (2\pi)^{\frac{1}{2}} R_\odot^3 \rho_{c\odot}^2 \frac{1}{z} \tag{3.59}$$

for large values of $d = z^2/T_c$, where z is given in (3.6). Inserting (3.58) into (3.56) we obtain finally the analytic representation of the relationship of the solar structure parameters and a solar neutrino emission rate for small values of the characteristic parameter $d = z^2/T_c$ in (3.57).

$$N_{12} \approx \left(1 - \frac{1}{2}\delta_{12}\right) 16\pi^{\frac{1}{2}} N_A^{\frac{5}{2}} \frac{(A_1 + A_2)^{\frac{1}{2}}}{(A_1 A_2)^{3/2}} S_{12}(0) X_1 X_2 \frac{R_\odot^3 \rho_{c\odot}^2}{T_{c\odot}^{1/2}}. \tag{3.60}$$

Inserting (3.59) into (3.56) we get finally the neutrino emission rate for large values of the characteristic parameter $d = z^2/T_c$ in (3.57) (Fig. 3.1).

$$N_{12} \approx \left(1 - \frac{1}{2}\delta_{12}\right) \left(\frac{2^7}{\pi}\right)^{\frac{1}{2}} \frac{N_A^3 h}{e^2} \frac{(A_1 + A_2)}{(A_1 A_2)^2 Z_1 Z_2} S_{12}(0) X_1 X_2 R_\odot^3 \rho_{c\odot}^2. \tag{3.61}$$

Fig. 3.1 Astrophysical Observatory Potsdam (AOP), Germany: Preparations for this book took into account the fact that the AOP celebrated its 150th Anniversary of the establishment of the Astrophysikalisches Observatorium Potsdam on 1st July 1874. AOP was the world's first observatory to emphasize explicitly the research area of astrophysics. Albert A. Michelson, studying under Hermann von Helmholtz at the Berlin University, developed his interferometer and performed the first time the famous Michelson experiment at AOP (cellar of the east dome, right hand-side of the photo) in 1881 supported by the then director of AOP, Herman C. Vogel. The authors are taking the opportunity to thank Dr. Reiner John (1942–2007, Astronomical Observatory Potsdam-Babelsberg) and Dr. Ewald Gerth (1934–2022, Astrophysical Observatory Potsdam-Telegrafenberg) for many years of exciting and productive cooperation on the issues of nuclear and neutrino astrophysics as described in this book. https://www.scirp.org/journal/paperinformation?paperid=134139

References

Bahcall, J.N.: Astrophys. J. **143**, 259–261 (1966)
Bahcall, J.N., Davis, Jr., R.: Essays in Nuclear Astrophysics. In: Barnes, C.A., Clayton, D.D., Schramm, D.N. (eds.) W.A. Fowler on the Occasion of his Seventieth Birthday, pp. 243–285. Cambridge University Press, Cambridge-London-New York-New Rochelle-Melbourne-Sydney (1982)
Bahcall, J.N., Huebner, W.F., Lubow, S.H., Parker, P.D., Ulrich, R.K.: Rev. Mod. Phys. **54**, 767–799 (1982)
Chandrasekhar, S.: An Introduction to the Study of Stellar Structure. Dover Publications, Inc., New York (1967). Unabridged, corrected republication of the 1st edn., University of Chicago Press, Chicago (1939)
Critchfield, C.L.: Cosmology, Fusion and Other Matters: George Gamow Memorial Volume. In: Reines, F. (ed.), pp. 186–191. Colorado Associated University Press, Colorado (1972)
Davis, Jr., R., Cleveland, B., Rowley, J.K.: Report on the solar neutrino experiments, presented at the conference on the intersections between particle and nuclear physics. Steamboat Springs, CO, May 23–30 (1984)
Fowler, W.A.: Unification of elementary forces and gauge theories. In: Cline, D.B., MIles, F.E. (eds.) Ben Lee Memorial International Conference On Parity Nonconservation, Weak Neutral Currents and Gauge Theories. Fermi National Accelerator Laboratory, Batavia, Illinois, U.S.A., Oct 20–22 (1977). Harwood Academic Publishers Ltd., London-Chur, pp. 509–527 (1977)
Fowler, W.A.: Rev. Mod. Phys. **56**, 149–179 (1984)
Haubold, H.J., Gerth, E.: Astron. Nachr. **304**, 299–304 (1983)
Haubold, H.J., John, R.W.: Fundamental problems in the theory of stellar evolution. In: Sugimoto, D., Lamb, D.Q., Schramm, D.N. (eds.) IAU Symposium No. 93, Proceedings, p. 317. D. Reidel Publishing Company, Dordrecht (1981)
Haubold, H.J., Mathai, A.M.: Ann. Phys. (Leipzig) **41**, 372–379 (1984)
Haubold, H.J., Mathai, A.M.: Ann. Phys. (Leipzig) **41**, 380–396 (1984)
Haubold, H.J., Mathai, A.M.: Astron. Nachr. **307**, 9–12 (1986)
Haubold, H.J., Mathai, A.M.: Ann. Phys. (Leipzig) **44**, 103–116 (1987)
Haxton, W.C.: Neutrino '84. 11th International Conference on Neutrino Physics and Astrophysics, Kleinknecht, K., Paschos, E.A. (eds.) Nordkirchen near Dortmund, F.R.G., June 11–16 (1984); World Scientific Publishing Co Pvt. Ltd., Singapore, pp. 217–228 (1984)
Hayashi, C., Hōshi, Sugimoto, D.: Evolution of the stars. Progr. Theor. Phys. (Japan) Suppl. **22**, 1 (1962)
Kavanagh, R.W.: Essays in Nuclear Astrophysics. In: Barnes, C.A., Clayton, D.D., Schramm, D.N. (eds.) W.A. Fowler on the Occasion of his Seventieth Birthday, pp. 159–170. Cambridge University Press, Cambridge-London-New York-New Rochelle-Melbourne-Sydney (1982)
Mathai, A.M., Saxena, R.K.: Generalized Hypergeometric Functions with Applications in Statistics and Physical Sciences. Lecture Notes in Mathematics, vol. 348. Springer, Berlin-Heidelberg-New York (1973)
Mathai, A.M., Saxena, R.K.: The H-function with Applications in Statistics and Other Disciplines. Wiley, New Delhi (1978)
Parker, P.D., Bahcall, J.N., Fowler, W.A.: Astrophys. J. **139**, 602–621 (1964)
Salpeter, E.E.: Phys. Rev. **88**, 547–553 (1952)
Saxena, R. K.: Proc. Nat. Acad. Sci. India **A26**, 400–413 (1960)

Open Access This chapter is licensed under the terms of the Creative Commons Attribution 4.0 International License (http://creativecommons.org/licenses/by/4.0/), which permits use, sharing, adaptation, distribution and reproduction in any medium or format, as long as you give appropriate credit to the original author(s) and the source, provide a link to the Creative Commons license and indicate if changes were made.

The images or other third party material in this chapter are included in the chapter's Creative Commons license, unless indicated otherwise in a credit line to the material. If material is not included in the chapter's Creative Commons license and your intended use is not permitted by statutory regulation or exceeds the permitted use, you will need to obtain permission directly from the copyright holder.

Chapter 4
Solar Nuclear and Neutrino Astrophysics Research, a Time Line

1930 Wolfgang Pauli hypothesizes the existence of neutrinos to account for the beta decay energy conservation crisis.

1933 Enrico Fermi writes down the correct theory for beta decay, incorporating the neutrino.

1956 Fred Reines and Clyde Cowan discover (electron anti-) neutrinos using a nuclear reactor.

The possible existence of neutrino oscillations, which is a consequence of neutrino masses and mixing, would be experimental evidence of elementary particle physics beyond the standard model.

1957 Bruno Pontecorvo proposes neutrino-antineutrino oscillations and this is the first time neutrino oscillations are hypothesized. B. Pontecorvo, Neutrino experiments and the problem of conservation of leptonic charge, Sov. Phys. JETP 26, 984 (1968).

1962 Ziro Maki, Masami Nakagawa and Sakata introduce neutrino flavour mixing and flavour oscillations. Z. Maki, M. Nakagawa, and S. Sakata, Remarks on the unified model of elementary particles, Prog. Theor. Phys. 28, 870 (1962).

1962 Muon neutrinos are discovered by Leon Lederman, Mel Schwartz, Jack Steinberger and colleagues at Brookhaven National Laboratories, and it is confirmed that they are different from electron neutrinos.

In the 1960s, the first experiment to detect solar neutrinos was Raymond Davis's Homestake Experiment, in which he observed a deficit in the flux of solar neutrinos with respect to the prediction of the Standard Solar Model, using a chlorine-based detector. In detecting solar neutrinos in the Homestake Experiment, it became clear that the number detected was much lower than that predicted by models of the solar interior. The problem could be solved by revising the model for the internal structure of the Sun (solar physics), the assumed mechanisms of thermonuclear reactions (nuclear physics), or the properties of neutrinos and understanding the limits of the detection mechanisms (neutrino physics). Most neutrinos passing through the Earth emanate from the Sun. About 65 billion solar neutrinos per second pass through every square centimetre perpendicular to the direction of the Sun in the region of the Earth.

Observations of solar neutrinos were first made by the Homestake experiment using a radiochemical method, and then followed by real-time measurement with KAMIOKANDE-II and other radiochemical experiments using gallium by SAGE and GALLEX/GNO:

R. Davis, Jr., D. S. Harmer, and K. C. Hoffman, Search for neutrinos from the Sun, Phys. Rev. Lett. 20, 1205 (1968).
K. S. Hirata, T. Kajita, T. Kifune, K. Kihara, M. Nakahata, K. Nakamura et al. (The Kamiokande-II Collaboration), Observation of ^8B solar neutrinos in the Kamiokande-II detector, Phys. Rev. Lett. 63, 16 (1989).
A. I. Abazov, O. L. Anosov, E. L. Faizov, V. N. Gavrin, A. V. Kalikhov, T. V. Knodel et al., Search for neutrinos from Sun using the reaction ^{71}Ga (electron-neutrino e-) ^{71}Ge, Phys. Rev. Lett. 67, 3332 (1991).
P. Anselmann et al. (The GALLEX Collaboration), Solar neutrinos observed by GALLEX at Gran Sasso, Phys. Lett. B 285, 376 (1992).
M. Altmann et al. (The GNO Collaboration), GNO solar neutrino observations: Results for GNO I, Phys. Lett. B 490, 16 (2000).

1963 R. F. Stein and A. G. W. Cameron (Eds.), Stellar Evolution, Proceedings of an international conference, November 13–15, 1963, Plenum Press 1966.

1964 John Bahcall and Raymond Davis discuss the feasibility of measuring neutrinos from the Sun and made the case for the Homestake Mine experiment.

1968 Raymond Davis and colleagues get first radiochemical solar neutrino results using cleaning fluid in the Homestake Mine in North Dakota, leading to the observed deficit subsequently known as the "solar neutrino problem".

1972 F. Reines (Ed.), Cosmology, Fusion @ Other Matters, George Gamow Memorial Volume, Colorado Associated University Press 1972.

1973 A. M. Mathai and R. K. Saxena, Generalized Hypergeometric Functions with Applications in Statistics and Physical Sciences, Lecture Notes in Mathematics, Vol. 348, Springer, Berlin-Heidelberg-New York 1973.

In 1974, conversation of HJH with Hans-Juergen Treder makes us aware that the solar neutrino problem is crucial for discovering new physics. Treder's advice was

McGill University

Department of Mathematics
Burnside Hall

Montreal
2 September 1982

Dr. H.J. Haubold
Zentralinstitut für Astrophysik
Akademie der Wisseschaften der DDR
DDR- 1500 Potsdam, Telegrafenberg
German Democratic Republic

Dear Dr. Haubold,

Thanks for your letter dated 14 August 1982 which reached me here just now. I am enclosing two copies of the corrected bibliography of our book.

I do not have ready references for integrals involving functions of the type $G_{p,0}^{0,p}$. One way of takling this type of integral is to replace the G-function by its Mellin - Barnes representation and then interchange the integral. The conditions are to be checked. In our book we did not work out the conditions in detail. Luke :The Special Functions and Their Approximations:, (Academic Press) Vol.I & II have lots of conditions worked out in detail.

Another way of tackling is to expand the gammas in the G-function by using any one of the asymptotic expansions and then integrate term by term which will also yield computable series representation for the integral.

The postal address of our university has changed. Please use the address given below for fast delivery of mail coming to our university.

If you would like to visit our university and spend a few days here then I can arrange some funds to meet part of your expenses. If you are interested please let me know about the items that you will need such as letters etc from me.

With best wishes
yours sincerely

Dr. A.M. Mathai
Professor

Postal address: 805 Sherbrooke Street West, Montreal, PQ, Canada H3A 2K6

Fig. 4.1 https://www.growkudos.com/projects/a-m-mathai-centre-for-mathematical-and-statistical-sciences-nurturing-the-love-for-mathematics

simple: "Ich glaube, dass die Werke der grossen Meister eine staendige Quelle von Ideen und neuen Zielstellungen fuer die aktuelle wissenschaftliche Arbeit sind". A similar advice was ones given by Albert Einstein (Figs. 4.1, 4.2 and 4.3).

With this advice in mind the topic solar neutrino radiation made us turning to the papers of the Proceedings of the first Solvay Council, held in 1911 focusing on The

CALIFORNIA INSTITUTE OF TECHNOLOGY

PASADENA, CALIFORNIA 91125

W. K. KELLOGG RADIATION LABORATORY 106-38

TELEPHONE (213) 795-6811

January 3, 1979

ZENTRA...
für As...
Eing.: 21. JAN 1979
Eing.-Nr. ...130...

Dr. H. J. Haubold
Zentralinstitut fur Astrophysik der Akademie der
    Wissenschatten der DDR
Potsdam, East Germany (DDR)

Dear Dr. Haubold,

    Thank you for sending me the corrigendum to your paper with Dr. R. W. John. It has occurred to me that in any future work you may wish to refer to C. L. Critchfield's Chapter 11, pg. 186 in Cosmology, Fusion & Other Matters, Editor: F. Reines, Colorado Associated University Press 1972. Critchfield's Eq. (17), corrected for typographical errors is commonly used in many of our calculations when his ξ (our τ) drops as low as unity but this rarely occurs.

                         Sincerely yours,

                         William A. Fowler
                         Institute Professor of Physics

WAF:mb

Fig. 4.2 https://www.nobelprize.org/prizes/physics/1983/fowler/facts/

Theory of Radiation and Quantum. Planck's quantum of action was discovered earlier but quantum mechanics not yet developed, and many questions discussed during the Council were destined to lead to new fundamental physics (or mathematics). Among them Einstein's comment that neither Herr Boltzmann nor Herr Planck has given a definition of W (probability in Boltzmann's entropy $S = k \ln W$) and Poincare's closing question if it was still possible to represent basic physical laws in terms of ordinary differential equations. We got stack at this point comparing physics of solar neutrino radiation and physics of black body radiation because we did not have an idea of Planck's "Staubkoernchen" (Pascual Jordan) for neutrino radiation.

1975 A. M. Mathai and P. N. Rathie, Basic Concepts in Information Theory and Statistics: Axiomatic Foundations and Applications, John Wiley & Sons, New Delhi 1975.

Already since 1975 we pursued the analysis of Homestake experiment data by Fourier analysis, wavelet analysis, and Lomb-Spargel periodograms prospectively to discover possible periodic variations in the publicly available data sets of the experiment. This analysis was jointly undertaken in cooperation with Kunitomo Sakurai (Kanagawa University, Yokohama, Japan). We shared the belief that the discovery of periodic variation of the solar neutrino flux may lead to new solar physics or new neutrino physics.

A summery of basic knowledge about solar physics, nuclear physics, and neutrino physics, having the Solar Neutrino Problem in mind, was provided in the first 18 pages of the 1988 edition of our book titled "Modern Problems in Nuclear and Neutrino Astrophysics". Back in 1974 we hoped that such a survey may provide us a way to contribute to the solution of the Solar Neutrino Problem.

May 22, 1985

Dear Dr. Haubold:

Thank you for your recent letter and for the reprints. The preprint for "Fortschritte" has not arrived at this time.

The only "simple" derivatives I have found for bipolar functions are those in eqs (9.9) and (9.10) of my lecture notes. In generic terms

$$(P'u)^2 = [P^2u + (psK_q + psK_1)^2][P^2u + (psK_q - psK_1)^2]$$

where p, q, r is any permutation of c, d, n.

I am enclosing copies of page-proofs that I have of Gamow's autobiography My World Line which Viking Press published in 1970. They tell about the early days of nuclear astrophysics.

In 1938 Bengt Strömgren was at the Carnegie Institution of Washington (Department of Terrestrial Magnetism) and they and the George Washington University devoted their annual conference on theoretical physics to stellar problems. As an authority on nuclear physics Bethe was invited to attend although Hans did not take astrophysics seriously at that time. Some months prior to the meeting I (as graduate student at G.W.U.) had suggested that a deuteron could be formed in the collision between protons in stars in accordance with the Gamow-Teller selection rule for beta-decay. They (Gamow and Teller were my supervisors) told me to forget it because von

(over)

Fig. 4.3 https://en.wikipedia.org/wiki/Charles_Critchfield

Weizsäcker had decided that the idea was not worth pursuing.

During the conference Bethe made that same suggestion whereupon Gamow and Teller told him that they had discouraged me from working on it. Hans then proposed that he and I develop the theory and publish together. That we did, independently, he in Cornell and I in Washington. A few months later (after conference) Hans wrote to us saying that he thought we should collaborate only on the p-p reaction. We could discern that he had discovered something, although he did not tell us what it was. It was the carbon-cycle, of course. Our results on the p-p reaction agreed, even to the detail that we both forgot the factor two that allows for either proton to become a neutron. After publication Robert Oppenheimer wrote to Hans reminding him that we had ignored the "matrix element of isotopic spin". I hope you will find these notes of interest.

Currently I am writing up my notes on elliptic functions and extending them into the transformation theory (and perhaps to elliptic modular functions) with the thought of possible publication. The theorists at LANL have urged me to do that and I enjoy it.

With very best regards,

Sincerely yours,
Charles L. Critchfield

Fig. 4.3 (continued)

1978 A. M. Mathai and R. K. Saxena, The H-function with Applications in Statistics and Other Disciplines, John Wiley and Sons, New Delhi 1978.

1979 we initiated an exchange of information with Raymond Davis Brookhaven National Laboratory asking him to provide us on a regular basis the data from his Homestake chlorine solar neutrino experiment. This cooperation by exchange of

BROOKHAVEN NATIONAL LABORATORY
ASSOCIATED UNIVERSITIES, INC.

Upton, Long Island, New York 11973

Department of Chemistry

(516) 282- 4322
FTS 666-

March 17, 1983

Dr. H. J. Haubold
Zentralinstitut für Astrophysik
DDR-1500
Potsdam, East Germany

Dear Professor Haubold:

Sorry for the long delay in answering your request for data from our chlorine solar neutrino experiment. I have enclosed copies of our latest reports given at meetings. In addition I am including a copy of our latest data plot and table. The columns in the data Table have the following meanings.

1. Date of the beginning of the exposure.
2. Date of the end of the exposure.
3. The date corresponding to the time at which one-half of the atoms that are collected would be produced assuming a uniform production rate.
4. The most likely value of the ^{37}Ar production rate in 615 metric tons of C_2Cl_4 in atoms per day.
5. The lower limit for the ^{37}Ar production rate in the same units, corresponds to a 1σ error.
6. The upper limit for the ^{37}Ar production rate in the same units, corresponding to a 1σ error.

This table includes all of the data analyzed. The remainder of the experimental runs are still being counted, so I do not have the results for 1982. Above columns 4, 5, and 6 are the combined values, however this average does not include runs nos. 71 and 72. However if these are included there will be very little difference.

If I can be of further help please let me know. Values for 1982 will come out slowly as the counting measurements are completed. If you are interested in the 1982 results please send me a note later on.

Sincerely yours,

Raymond Davis Jr.

jd
enc.

Fig. 4.4 https://www.nobelprize.org/prizes/physics/2002/davis/biographical/

letters and visits to Raymond Davis workplace at the University of Pennsylvania was active until 2002 (Figs. 4.4 and 4.5).

At the same time, we also took up a problem that was discussed in the 1970s concerning the closed-form evaluation of thermonuclear reaction rates described in publications of Bethe, Salpeter, and Fowler and explicitly discussed by Charles Critchfield. Critchfield's paper can be considered, in principle, as a follow-up research result to

KANAGAWA UNIVERSITY
INSTITUTE OF PHYSICS
ROKKAKUBASHI, YOKOHAMA, 221, JAPAN

September 30, 1989

Prof. H.J. Haubold
Room S-3260B
Outer Space Division
United Nations
New York, N.Y. 10017
U.S.A.

Dear Professor Haubold:

It is nice to receive a letter from you I have known for many years since our first meeting in India.

In the letter not dated, you request a copy of my paper entitled "Solar Neutrino Problem as Viewed from the Active Phenomena on the Sun." This is printed in the Proceedings of XVI-th INS Internatl. Symp., held in March 1988. No reprint was made of it, so herewith enclosed is the Xeroxed copy taken from the proceedings. Also enclosed herewith is a sheet of the notice for our new books on cosmic ray astrophysics and neutrino astrophysics. I guess you are interested to see them.

I am now planning to attend the 21st ICRC to be held in Australia in this coming January and to present a paper on the solar neutrinos, in which it is shown that a chaotic process is taking place inside the sun.

I am looking forward hearing from you about your current status some day.

With best wishes.

Kunitomo Sakurai
Professor of Physics

Fig. 4.5 https://www.walshmedicalmedia.com/open-access/kunitomo-sakurai-solar-neutrino-problems-how-they-were-solved-2332-2519-1000132.pdf

his paper published in cooperation with Hans Bethe in 1939 that led to the Nobel Prize for Bethe in in 1967 for his work on the production of energy in stars.

1980s The Kamioka experiment is built in a zinc mine in Japan.

1983 William A. Fowler receives the Nobel Prize in Physics for his theoretical and experimental studies of the nuclear reactions of importance in the formation of the chemical elements in the universe: W.A. Fowler, Experimental and theoretical nuclear astrpphysics: the quest for the origin of the elements, Reviewes of Modern Physics 56 (1984) 149–179.

1985 The "atmospheric neutrino anomaly" is observed by IMB and Kamiokande.

1986 T. Pinch, Confronting Nature: The Sociology of Solar-Neutrino Detection, D. Reidel Publishing CompanyDordrecht.

1986 Kamiokande group makes first directional counting observation solar of solar neutrinos and confirms deficit.

1987 The Kamiokande and IMB experiments detect burst of neutrinos from Supernova 1987A in the Large Magellanic Cloud, a satellite galaxy of the Milky Way, heralding the birth of neutrino astronomy, and setting many limits on neutrino properties, such as mass.

1987 conversation of HJH with Hans A. Bethe and F. Reines, meeting at the Michelson-Morley Conference 1987 in Cleveland. Our research work started focusing on the Big Bang model, including Big Bang nucleosynthesis and predictions of the cosmic microwave background radiation.

S. Gottloeber, HJH, J. P. Muecket, and V. Mueller, Early Evolution of the Universe and Formation of Structure, Akademie-Verlag, Berlin 1990, de Gruyter 2024).

The following is the generalized form of the basic reaction-rate probability integral in the real scalar positive variable case:

$$I_{a,b}(\delta, \rho) = \int_0^\infty v^{\gamma-1} e^{-av^\delta - bv^{-\rho}} dv$$

for $a > 0, b > 0, \delta > 0, \rho > 0, \Re(\gamma) > 0$. Before the integral is evaluated, let us consider some particular cases. For $\delta = 1, \rho = 1$, it is the basic Bessel integral. For $\delta = 1, \rho = 1$, the integrand, normalized, is the inverse Gaussian density. For $\delta = 1, \rho = \frac{1}{2}$, it is the reaction-rate probability integral in nuclear reaction-rate theory. For $\delta = 1$ and general ρ, it is Krätzel integral and Krätzel transform is associated with it. The integral is also known in the literature by different names such as generalized gamma integral, ultra gamma integral and super gamma integral.

The closed-form representation of the integral can be written in terms of a H-function, namely,

$$g(u) = \frac{1}{\rho \delta a^{\gamma/\delta}} H_{0,2}^{2,0}\left[a^{\frac{1}{\delta}} u \Big|_{(0,\frac{1}{\rho}),(\frac{\gamma}{\delta},\frac{1}{\delta})}\right], u = b^{\frac{1}{\rho}}$$

where the c in the contour is such that $c > 0, \delta > 0, \rho > 0, \Re(\gamma) > 0$.

1991 SAGE (in Russia) and GALLEX (in Italy) confirm the solar neutrino deficit in radiochemical experiments.

1995 H. J. Haubold and A. M. Mathai, A Heuristic Remark on the Periodic Variation in the Number of Solar Neutrinos Detected on Earth, Astrophysics and Space Science 228, 113–134.

1995 With Reiner John, Ewald Gerth, under the guidance of Arak M. Mathai, we stopped analyzing the data of Davis's Homestake experiment and turned our attention to the operation of the SuperKamiokande experiment and the analysis of publicly made available solar neutrino data.

1996 SuperKamiokande, the largest particle detector ever, begins searching for neutrino interactions on 1 April at the site of the Kamioka experiment, with a Japan-US team.

1998 Analysing more than 500 d of data, the SuperKamiokande team reports evidence of oscillations in atmospheric neutrinos implying that neutrinos have nonzero mass, thus suggesting physics beyond the Standard Model of Particle Physics.

2001 The Sudbury Neutrino Observatory (SNO) reported observation of neutral currents from solar neutrinos, along with charged currents and elastic scatters, providing convincing evidence that neutrino oscillations are the cause of the solar neutrino deficit.

An initial indication of solar neutrino oscillations was obtained from the difference between the 8B solar neutrino fluxes as measured in the elastic-scattering channel at SuperKamiokande and the charged-current channel at the Sudbury Neutrino Observatory in 2001. Solar neutrino oscillation was subsequently established by including neutral-current measurements from SNO. Solar neutrino oscillations were confirmed using reactor antineutrinos by KamLAND (Fig. 4.6):

S. Fukuda et al. (The SuperKamiokande Collaboration), Solar ^8B and hep neutrino measurements from 1258 d of SuperKamiokande data, Phys. Rev. Lett. 86, 5651 (2001).
Q. R. Ahmad et al. (The SNO Collaboration), Measurement of the rate of $\nu_e + d \to p + p + e^-$ interactions produced by ^8B solar neutrinos at the Sudbury Neutrino Observatory, Phys. Rev. Lett. 87, 071301 (2001).
Q. R. Ahmad et al. (The SNO Collaboration), Direct evidence for neutrino flavor transformation from neutral current interactions in the Sudbury Neutrino Observatory, Phys. Rev. Lett. 89, 011301 (2002).
K. Eguchi et al. (The KamLAND Collaboration), First results from Kamland: Evidence for reactor antineutrino disappearance, Phys. Rev. Lett. 90, 021802 (2003).

2002 Raymond Davis and Masatoshi Koshiba won shares of the Nobel Prize in Physics for their roles in the detection of neutrinos from the Sun and Supernova 1987A: R. Davis, Jr., A half-century with solar neutrinos, Reviewes of Modern Physics 75 (2003) 985–994.

2010 N. Scafetta, Fractal and Diffusion Entropy Analysis of Time Series: Theory, concepts, applications and computer codes for studying fractal noises and Levy Walk signals, VDM Verlag Dr. Mueller, Saarbruecken 2010.

2014 Analysis of Solar Neutrino Data from SuperKamiokande I and II, Entropy 16, 1414–1425.

In physics one is used to be confronted with either difficult experimental results or ambitious theoretical considerations in one way or another. The time series of data from solar neutrino experiments like Homestake (US) and SuperKamiokande (Japan) appear to be noisy and irregular. The authors have analysed solar neutrino data from the two experiments since 1974 by applying mathematically rigorous

```
November 27, 1989

Dr. Hans Joachim Haubold
United Nations
United Nations, NY 10017

Dear Dr. Haubold:

This is in response to your letter of November 14, 1989. I am
sorry to have been so long in responding but I have been away.

Since you have seen our first chapter in the Gamow memorial
volume, you are already aware that Gamow made a very simple
calculation, using the Jeans' criterion, to obtain a value for
the diameter of a typical galaxy based on fundamental constants.
Robert Herman and I corrected some minor errors in his
formulation which we also published in Nature, and which is
referred to in the memorial volume.

We later attempted to carry the discussion based on the Jeans'
criterion a bit further in a paper we wrote in 1967, but which we
never published. Perhaps it will be of some interest to you, and
I therefore enclose a copy.

Considerations such as your mass scale formula have been made by
several investigators, starting, I believe, with Chandrasekhar,
as early as 1937. You will find such matters quite adequately
reviewed in the book The Anthropic Cosmological Principle by
Barrow and Tipler (Oxford University Press 1988 - paperback
edition). On page 234 a mass scale formula is given with the
exponent zeta instead of your exponent (n/2). This formula does
give a typical mass figure for a protogalaxy.

If you have read my paper with Gamow on the fundamental constants
or my American Scientist article on the subject, you will realize
that I believe the Dirac LNH is really based on dimensional
considerations in which he used the age of the universe as a
characteristic time. It follows then that to form dimensionless
ratios one must have some other dimensional constant also vary
with the epoch. We now know that the variation of G with the
epoch is observationally unacceptable (work of Canuto, Goldberg,
Shapiro, etc.). One can use other characteristic times, such as
the time of decoupling, which is a number unique to a particular
cosmological model. This yields Dirac-like ratios, without the
problem of a dependence of anything on the epoch.

I have no problem in using the results of dimensional analysis in
the absence of a theory, but feel that one has to be careful not
to read too much into the results. For example, over the years
people seem to have forgotten the origin of the Planck time, so
that it achieves enormous importance in analyses of the early
universe.

I do not want to analyze the reprint you sent me in detail, but
```

Fig. 4.6 https://en.wikipedia.org/wiki/Ralph_Alpher

```
do note with some surprise that you have not used the
temperature-time relation determined by the dominant radiation
field, which seems to me to be appropriate for the temperature in
the expanding universe during the time prior to radiation
decoupling. (See the paper by Alpher and Herman in Physics
Today.) This will surely affect your result.

Sincerely,
```

Ralph A. Alpher

```
Ralph A. Alpher
Department of Physics
Union College
Schenectady, NY 12308


CC:Dr. Robert Herman
```

Fig. 4.6 (continued)

Fourier analysis, wavelet analysis, Lomb-Scargle periodograms. In the case of the SuperKamiokande solar neutrino data we made the decision to study the data with Standard Deviation Analysis (SDA) and Diffusion Entropy Analysis (DEA). The purpose is to study the scaling exponent of a complex time series that may manifest long-range correlations and fractal statistics. Scaling analysis (SDA, Hurst analysis, Detrended Fluctuation Analysis, etc.) relays on the assumption that physical data are characterised by fractal Brownian memory. Such methods of scaling analysis are based on the evaluation of the variance of a diffusion process. DEA does NOT relay on the assumption of fractal Brownian memory. DEA evaluates the scaling exponent of the probability density function (pdf) through the Shannon entropy of the diffusion process generated by those fluctuating data. SDA ad similar methodologies evaluate only the scaling of the variance and NOT the pdf scaling. One can call "H" (Hurst) the scaling exponent detected by means of the variance-based methods (SDA) and "delta" the scaling exponent detected by DEA. Our results show that "H" and "delta" are not identical. This means that the SuperKamiokande data are not characterized by fractal Gaussian statistics. It further means that the data are characterized by Levy statistics. Additionally, we found that "delta" < 1 which clearly indicated Levy flights and super-diffusion. This research result, by analysing SuperKamiokande data, did even allow us to identify the fractal differential equation that is guiding the now called solar neutrino probability density function (pdf). This pdf we identified in terms of a Fox H-function H[1,1;2,2]. The solar neutrino pdf was discovered by analysing SuperKamiokande experiment data. The opposite (and most welcome!) approach to the SuperKamiokande experiment has been published in papers titled

Probability Density Function for Neutrino Masses and Mixings, Phys. Rev. D94, 115004 (2016) and
Mellin Transform Approach to Rephasing Invariants, Phys. Rev. D102, 036001 (2020).

The authors discover the solar neutrino probability density in terms of Fox' H-function by assuming the probability density function:

$$p(x,t) = \frac{1}{t^\delta} f\left(\frac{x}{t^\delta}\right), \qquad (4.1)$$

where δ denotes the scaling exponent of the pdf. In the variance based methods, scaling is studied by direct evaluation of the time behavior of the variance of the diffusion process. If the variance scales, one would have

$$\sigma_x^2(t) \sim t^{2H}, \qquad (4.2)$$

where $\sigma_x^2(t)$ is the variance of the diffusion process and where H is the Hurst exponent. To evaluate the Shannon entropy of the diffusion process at time t, defined $S(t)$ as

$$S(t) = -\int_{-\infty}^{+\infty} dx\, p(x,t) \ln p(x,t) \qquad (4.3)$$

and with the previous $p(x,t)$ one has
The solution of the generalized diffusion equation

$$\frac{\partial p(x,t)}{\partial t} = K^\alpha {}_{-\infty}D_x^\alpha p(x,t), \qquad (4.4)$$

where ${}_{-\infty}D_x^\alpha$ is the fractional Weyl operator, is

$$p(x,t) = \frac{1}{\alpha} \frac{1}{t^{1/\alpha}} \frac{t^{1/\alpha}}{|x|} H_{2,2}^{1,1}\left[\frac{|x|}{Kt^{1/\alpha}} \Big|_{(1,1),(1,\frac{1}{2})}^{(1,\frac{1}{\alpha}),(1,\frac{1}{2})}\right]$$

$$= \frac{1}{\alpha|x|} \frac{1}{2\pi i} \int_{c-i\infty}^{c+i\infty} \frac{\Gamma(1+\frac{s}{\alpha})\Gamma(-\frac{s}{2})}{\Gamma(-s)\Gamma(1+\frac{s}{2})} \left(\frac{|x|}{Kt^{1/\alpha}}\right)^{-s} ds$$

2018 A. M. Mathai and H. J. Haubold, Erdelyi-Kober Fractional Calculus: From a Statistical Perspective, Inspired by Solar Neutrino Physics, Springer Briefs in Mathematical Physics 31, Springer, Singapore.

Since these discoveries, Borexino and KamLAND experiments have measured the neutrino fluxes from different solar nuclear fusion processes, such as pp, pep, ^7Be, and Carbon-Nitrogen-Oxygen cycle:
M. Agostini et al. (The Borexino Collaboration), First simultaneous precision spectroscopy of pp, ^7Be, and pep solar neutrinos with Borexino phase-II, Phys. Rev. D 100, 082004 (2019).
M. Agostini et al. (The BOREXINO Collaboration), Comprehensive measurement of pp-chain solar neutrinos, Nature 562, 505 (2018).
A. Gando et al. (The KamLAND Collaboration), ^7Be solar neutrino measurement with KamLAND, Phys. Rev. C 92, 055808 (2015).

M. Agostini et al. (The BOREXINO Collaboration), Experimental evidence of neutrinos produced in the CNO fusion cycle in the Sun, Nature (London) 587, 577 (2020).

All measurements to date are naturally explained by neutrino flavor change due to neutrino oscillations with matter effects predicted by Mikheyev, Smirnov, and Wolfenstein, termed the MSW effect: higher energy neutrinos undergo adiabatic conversion from the electron flavor state to the second mass eigenstate. While neutrino oscillations and MSW effect is consistent with all current solar neutrino measurements, two distinctive predictions are yet to be observed: the characteristic energy dependence of the solar neutrino electron-flavor survival probability $P_{ee}(E_\nu)$ distortion due to the MSW effect in the Sun and the day/night flux asymmetry induced by the matter effect in the Earth. One of the interests of solar neutrino experiments is to determine the neutrino oscillation parameters of Δm_{21}^2 and $\sin^2 \Delta_{12}$. Independent of solar neutrino measurements, the KamLAND experiment used reactor anti-neutrinos to measure the same oscillation parameters, assuming CPT symmetry holds:

S. P. Mikheyev and A. Y. Smirnov, Resonance amplification of oscillations in matter and spectroscopy of solar neutrinos, Sov. J. Nucl. Phys. 42, 913 (1985).
L. Wolfenstein, Neutrino oscillations in matter, Phys. Rev. D 17, 2369 (1978).
A. J. Baltz and J. Weneser, Effect of transmission through the Earth on neutrino oscillations, Phys. Rev. D 35, 528 (1987).
J. Bouchez, M. Cribier, J. Rich, M. Spiro, D. Vignaud, and W. Hampel, Matter effects for solar neutrino oscillations, Z. Phys. C 32, 499 (1986).
E. D. Carlson, Terrestrially enhanced neutrino oscillations, Phys. Rev. D 34, 1454 (1986).
M. Cribier, W. Hampel, J. Rich, and D. Vignaud, MSW regeneration of solar ν_e in the Earth, Phys. Lett. B 182, 89 (1986).
S. T. Petcov, Diffractive like (or parametric resonance like?) enhancement of the Earth (day night) effect for solar neutrinos crossing the Earth core, Phys. Lett. B 434, 321 (1998).
P. Bakhti and A. Y. Smirnov, Oscillation tomography of the Earth with solar neutrinos and future experiments, Phys. Rev. D 101, 123031 (2020).
A. Gando, Y. Gando, H. Hanakago, H. Ikeda, K. Inoue, K. Ishidoshiro et al. (The KamLAND Collaboration), Reactor on-off antineutrino measurement with KamLAND, Phys. Rev. D 88, 033001 (2013).

Many subsequent radiochemical and water Cherenkov detectors confirmed the deficit, but neutrino oscillation was not conclusively identified as the source of the deficit until the Sudbury Neutrino Observatory provided clear evidence of neutrino flavour change in 2001. Solar neutrinos have energies below 20 MeV and travel an astronomical unit between the source in the Sun and detector on the Earth. At energies above 5 MeV, solar neutrino oscillations actually take place in the Sun through a resonance known as the MSW effect, a different process from the vacuum oscillation. The transition between the low energy regime (the MSW effect is negligible) and the high energy regime (the oscillation probability is determined by matter effects) lies

4 Solar Nuclear and Neutrino Astrophysics ...

in the region of about 2 MeV for the solar neutrinos. The MSW effect is important at the very large electron densities of the Sun where electron neutrinos are produced. The high-energy neutrinos seen, for example, in Sudbury Neutrino Observatory and in SuperKamiokande, are produced mainly as the higher mass eigenstate in matter and remain as such as the density of solar material changes. When neutrinos go through the MSW resonance the neutrinos have the maximal probability to change their nature, but it happens that this probability is negligibly small this is sometimes called propagation in the adiabatic regime.

The Chap. 5 "Nuclear Astrophysics, 2025 Update", first, gives the explicit evaluation of the basic reaction-rate probability integral and its representations in computable series form. Then, a more general form of the reaction-rate probability integral, which can be called the generalized Bessel integral, is examined. Here also, computable series forms are given. Connection of this model to Krätzel integral, Krätzel transform, inverse Gaussian density etc. is established. Next, Mathai's pathway extension of the reaction-rate probability integral is explored. The discussion so far is for real scalar variable situations. A multivariate extension or a p-variate model for the generalized reaction-rate integral is explored. This is then extended to the corresponding integral in the complex domain. Next, a real $p \times q$ matrix-variate integral is discussed, which can be considered as the real matrix-variate extension of the reaction-rate probability integral. Then, the matrix-variate integral is extended to the complex domain. All the above topics can be considered as Mellin convolutions and M-convolutions of a product involving generalized gamma functions as the basic functions. Next, other functions are incorporated into a Mellin convolution of a product. It is shown that when a type-1 beta form of the function is considered, one can reach fractional integral of the second kind. This idea is extended to real and complex matrix-variate cases. Then, a Mellin convolution of a ratio is examined and its connections to fractional integral of the fist kind are established. These ideas are extended to matrix-variate cases in the real and complex domains. The purpose of this chapter is to discuss mathematical aspects connected with generalized reaction-rate probability integral so that physicists and others can explore the possibility of the corresponding new physics or communication theory.

The Chap. 6 "Neutrino Astrophysics, 2025 Update: The Entropic Approach to Solar Neutrinos" is utilizing data from the SuperKamiokande solar neutrino detection experiment and analyses them by diffusion entropy analysis and standard deviation analysis. The main result of analysis indicates that solar neutrinos are subject to Lévy flights (anomalous diffusion, super-diffusion). Subsequently the chapter derives the probability density function and the governing fractional diffusion equation (fractional Fokker-Planck Equation) for solar neutrino Lévy flights. The conclusion is Does SuperKamiokande Observe Lévy Flights of Solar Neutrinos?

The Chap. 7 "Neutrino Astrophysics, 2025 Update: Neutrino Masses and Mixings" is addressing one of the highest priorities in fundamental particle physics concerning the discovery of non-zero neutrino masses that remains one of the very few hints regarding the nature of physics beyond the Standard Model of Elementary Particle Interactions. The questions about neutrino properties still remain unanswered, such as the their absolute masses, the ordering of the mass states and the charge-parity violating phase. The neutrino sector of the seesaw-modified Standard Model is investigated under the anarchy principle. The anarchy principle leading to the seesaw ensemble is studied analytically with tools of random matrix theory. The probability density function is obtained (Fig. 4.7).

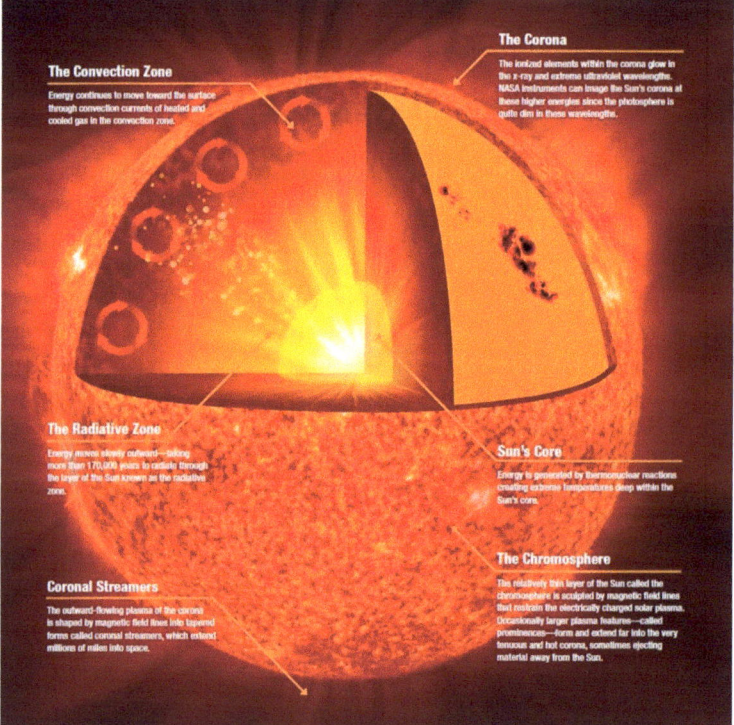

Fig. 4.7 Nuclear Astrophysics: The diagram shows the internal structure of the Sun. The interior of the Sun is a ball of swirling hot plasma that is held together by a balance of forces between gravity and pressure. The dense gases inside the Sun are so massive that they create a strong gravitational pull, which helps to keep solar material from escaping. As a counter force, the expanding hot gases create a large amount of pressure pushing outward toward the Sun's surface. The push and pull between gravity and pressure create conditions that maintain the three interior regions of the Sun: the core, the radiative zone, and the convective zone. The core is the centre of the Sun and extends about a quarter of the way to the surface. About half of the Sun's mass is within the core. Even though the core is made of gas, it is 10 times denser than lead. It is also the hottest region of the Sun, about 15 million degrees Celsius (27 million degrees Fahrenheit). The Sun's core is the only place in our solar system where the temperature and density conditions are high enough for nuclear fusion reaction to occur naturally. The nuclear reactions within the core fuse hydrogen atoms into helium atoms, releasing extremely large amounts of energy in the process. Some of the energy that is created in the core travels to the surface of the Sun through the Sun's atmosphere, and out into space, enough of it reaching Earth's surface to sustain life. An in terms of physics, very important by-product of the nuclear reactions are the neutrinos

Open Access This chapter is licensed under the terms of the Creative Commons Attribution 4.0 International License (http://creativecommons.org/licenses/by/4.0/), which permits use, sharing, adaptation, distribution and reproduction in any medium or format, as long as you give appropriate credit to the original author(s) and the source, provide a link to the Creative Commons license and indicate if changes were made.

The images or other third party material in this chapter are included in the chapter's Creative Commons license, unless indicated otherwise in a credit line to the material. If material is not included in the chapter's Creative Commons license and your intended use is not permitted by statutory regulation or exceeds the permitted use, you will need to obtain permission directly from the copyright holder.

Chapter 5
Nuclear Astrophysics, 2025 Update

5.1 Explicit Evaluation of the Thermonuclear Reaction-Rate Probability Integrals

For the sake of ready reference, the basic materials will be restated here. Let $x_1 > 0$ and $x_2 > 0$ be two real scalar positive variables with the associated functions $f_1(x_1)$ and $f_2(x_2)$ respectively. Let the joint function of x_1 and x_2 be $f_1(x_1) f_2(x_2)$, the product. If $x_1 > 0$ and $x_2 > 0$ are real scalar random variables with the densities $f_1(x_1)$ and $f_2(x_2)$, then we say that x_1 and x_2 are statistically independently distributed when we take the joint density as $f_1(x_1) f_2(x_2)$, the product. Let $u = x_1 x_2$ the product. Consider the transformation $u = x_1 x_2$, $v = x_2$. Then, we can see that the wedge product of differentials are connected by the relation $dx_1 \wedge dx_2 = \frac{1}{v} du \wedge dv$ and then the marginal function of u, denoted by $g(u)$, is given by the following:

$$g(u) = \int_0^\infty \frac{1}{v} f_1\left(\frac{u}{v}\right) f_2(v) dv = \int_0^\infty \frac{1}{v} f_1(v) f_2\left(\frac{u}{v}\right) dv \tag{5.1}$$

where the second form of the integral is obtained by taking the transformation as $u = x_1 x_2$ and $v = x_1$. Now, consider the evaluation of the following integral denoted by $I_{a,b}$:

$$I_{a,b} = \int_0^\infty v^{\gamma-1} e^{-av - bv^{-\frac{1}{2}}} dv. \tag{5.2}$$

This (5.2) is the basic reaction-rate probability integral in nuclear reaction-rate theory (Mathai and Haubold 1988). Let us denote the Mellin transform of a real-valued scalar function $f(x)$ of the real scalar variable x, with Mellin parameter s, as $M_f(s)$, that is, $M_f(s) = \int_0^\infty x^{s-1} f(x) dx$ whenever the integral is convergent. In order to evaluate the integral in (5.2) by using (5.1), let us take

$$f_1(x_1) = e^{-x_1^{\frac{1}{2}}}, x_1 > 0 \Rightarrow M_{f_1}(s) = \int_0^\infty x_1^{s-1} e^{-x_1^{\frac{1}{2}}} dx_1 = 2\Gamma(2s), \Re(s) > 0$$

where $\Re(\cdot)$ denotes the real part of $(\cdot)$, and let

$$f_2(x_2) = x_2^\gamma e^{-ax_2}, x_2 > 0, a > 0 \Rightarrow$$
$$M_{f_2}(s) = \int_0^\infty x_2^{\gamma+s-1} e^{-ax_2} dx_2 = \Gamma(\gamma+s) a^{-(\gamma+s)}, \Re(s) > -\Re(\gamma). \quad (5.3)$$

Note that

$$f_1\left(\frac{u}{v}\right) = e^{-(\frac{u}{v})^{\frac{1}{2}}} = e^{-bv^{-\frac{1}{2}}} \Rightarrow b = u^{\frac{1}{2}} \text{ or } u = b^2$$

and $\frac{1}{v} f_2(v) = \frac{1}{v} v^\gamma e^{-av} = v^{\gamma-1} e^{-av}$. Note that, by taking the Mellin transform of g in (5.1), we have $M_g(s) = M_{f_1}(s) M_{f_2}(s)$, which is the Mellin convolution of a product property also. This property is established easily.

$$M_g(s) = \int_0^\infty u^{s-1} g(u) du = \int_{u=0}^\infty \left[\int_{v=0}^\infty \frac{1}{v} f_1\left(\frac{u}{v}\right) f_2(v) dv\right] du$$
$$= \int_0^\infty \int_0^\infty x_1^{s-1} f_1(x_1) x_2^{s-1} f_2(x_2) dx_1 \wedge dx_2 = M_{f_1}(s) M_{f_2}(s) \quad (5.4)$$

where we have reused the same transformation $u = x_1 x_2$ with the associated Jacobian $dx_1 \wedge dx_2 = \frac{1}{v} du \wedge dv$ at the second line of the above derivation. Now, from (5.3) and (5.4), we have

$$M_g(s) = M_{f_1}(s) M_{f_2}(s) = 2\Gamma(2s) a^{-(\gamma+s)} \Gamma(\gamma+s), \Re(s) > 0, \Re(s) > -\Re(\gamma). \quad (5.5)$$

Then, by taking the inverse Mellin transform of (11.5) we have g. That is,

$$g(u) = \frac{1}{2\pi i} \int_{c-i\infty}^{c+i\infty} \frac{2}{a^\gamma} \Gamma(2s) \Gamma(\gamma+s) (au)^{-s} ds \quad (5.6)$$

where $i = \sqrt{(-1)}$ and c in the contour is any real number > 0. This, (5.6) can be written as a H-function. That is,

$$g(u) = \frac{2}{a^\gamma} H_{0,2}^{2,0}\left[au\big|_{(0,2),(\gamma,1)}\right], u \geq 0. \quad (5.7)$$

For the theory and applications of the H-function, see Mathai et al. (2010). By using the duplication formula for the gamma function, see for example, Mathai (1993), one can write

$$2\Gamma(2s) = \pi^{-\frac{1}{2}} 2^{2s} \Gamma(s) \Gamma\left(s + \frac{1}{2}\right). \quad (5.8)$$

5.1 Explicit Evaluation of the Thermonuclear Reaction-Rate Probability Integrals

Now, the integrand in (5.6) can be written as

$$\frac{1}{a^\gamma \pi^{\frac{1}{2}}} \Gamma(s) \Gamma\left(s + \frac{1}{2}\right) \Gamma(\gamma + s) \left(\frac{a}{4}u\right)^{-s}.$$

which enables us to write the H-function in terms of a G-function, that is,

$$g(u) = \frac{1}{a^\gamma \pi^{\frac{1}{2}}} G_{0,3}^{3,0}\left[\frac{au}{4} \Big|_{0,\frac{1}{2},\gamma}\right], u \geq 0. \tag{5.9}$$

For the theory and applications of the G-function, see, for example, Mathai (1993). Note that in the gamma product $\Gamma(s)\Gamma(s + \frac{1}{2})\Gamma(\gamma + s)$ the poles are simple as long as γ is not an integer or half-integer. In this case, one can write the G-function representation in (5.9) in terms of simple hypergeometric series. Note that the poles of $\Gamma(s)$ are at $s = -\nu, \nu = 0, 1, \cdots$. The poles of $\Gamma(\frac{1}{2} + s)$ are at $s = -\frac{1}{2} - \nu, \nu = 0, 1, \cdots$ and the poles of $\Gamma(\gamma + s)$ are at $s = -\gamma - \nu, \nu = 0, 1, \cdots$. Hence, for $\gamma \neq \frac{\nu}{2}, \nu = 0, 1, \cdots$ the sum of the residues at the poles of $\Gamma(s)$ is the following, denoted by S_1:

$$S_1 = \sum_{\nu=0}^{\infty} \frac{(-1)^\nu}{\nu!} \Gamma\left(\frac{1}{2} - \nu\right) \Gamma(\gamma - \nu) \left(\frac{a}{4}u\right)^\nu$$

$$= \Gamma\left(\frac{1}{2}\right) \Gamma(\gamma) \sum_{\nu=0}^{\infty} \frac{1}{(1-\gamma)_\nu (\frac{1}{2})_\nu} \left(\frac{a}{4}u\right)^\nu$$

$$= \Gamma\left(\frac{1}{2}\right) \Gamma(\gamma) {}_0F_2\left(\,; 1-\gamma, \frac{1}{2}; -\frac{a}{4}u\right), u = b^2, \tag{5.10}$$

where, for example, $(a)_m = a(a+1)\cdots(a+m-1), a \neq 0, (a)_0 = 1$ is the Pochhammer symbol. For writing (5.10) in terms of the hypergeometric series ${}_0F_2$, we have used basically two properties:

$$\lim_{z \to -\nu}(z+\nu)\Gamma(z) = \lim_{z \to -\nu}(z+\nu)\frac{(z+\nu-1)\cdots z}{(z+\nu-1)\cdots z}\Gamma(z)$$

$$= \lim_{z \to -\nu}\frac{\Gamma(z+\nu+1)}{[(z+\nu-1)\cdots z]} = \frac{(-1)^\nu}{\nu!}\Gamma(1) = \frac{(-1)^\nu}{\nu!}. \tag{5.11}$$

$$\Gamma(z) = (z-1)(z-2)\cdots(z-\nu)\Gamma(z-\nu) \Rightarrow$$

$$\Gamma(z - \nu) = \frac{\Gamma(z)}{(z-1)(z-2)\cdots(z-\nu)}$$

$$= \frac{(-1)^\nu}{(1-z)(1-z+1)\cdots(1-z+\nu-1)} = \frac{(-1)^\nu}{(1-z)_\nu} \tag{5.12}$$

whenever $\Gamma(z)$ and $\Gamma(z - \nu)$ are defined. Now, evaluating the sum of the residues at the poles of $\Gamma(s + \frac{1}{2})$, that is at $s = -\frac{1}{2} - \nu, \nu = 0, 1, \ldots$, we have the following, denoted by S_2:

$$S_2 = \sum_{\nu=0}^{\infty} \frac{(-1)^\nu}{\nu!} (au/4)^{\frac{1}{2}} \Gamma\left(-\frac{1}{2} - \nu\right) \Gamma\left(\gamma - \frac{1}{2} - \nu\right) (au/4)^\nu$$

$$= (au/4)^{\frac{1}{2}} \Gamma\left(-\frac{1}{2}\right) \Gamma(\gamma - \frac{1}{2})_0 F_2\left(\ ; \frac{3}{2}, \frac{3}{2} - \gamma; -au/4\right), u = b^2. \quad (5.13)$$

The sum of the residues at the poles of $\Gamma(s + \gamma)$ or at the points $s = -\gamma - \nu, \nu = 0, 1, \ldots$, is the following, denoted by S_3, by using steps parallel to the ones used in (5.10) and (5.13):

$$S_3 = \sum_{\nu=0}^{\infty} \frac{(-1)^\nu}{\nu!} \left(\frac{au}{4}\right)^\gamma \Gamma(-\gamma - \nu) \Gamma\left(\frac{1}{2} - \gamma - \nu\right) \left(\frac{au}{4}\right)^\nu$$

$$= \Gamma(-\gamma) \Gamma\left(\frac{1}{2} - \gamma\right) \left(\frac{au}{4}\right)^\gamma {}_0F_2\left(\ ; 1+\gamma, \frac{1}{2}+\gamma; -\frac{au}{4}\right), u = b^2. \quad (5.14)$$

From (5.10), (5.13) and (5.14), we have the following series form for $I_{a,b}$, which will be stated as a theorem.

Theorem 5.1 *For* $a > 0, b > 0, \Re(\gamma) > 0$, *the reaction-rate probability integral*

$$I_{a,b} = \int_0^\infty x^{\gamma-1} e^{-ax - bx^{-\frac{1}{2}}} \, dx = \frac{2}{a^\gamma} H_{0,2}^{2,0}\left[ab^2 \big|_{(0,2),(\gamma,1)}\right], u \geq 0$$

$$= a^{-\gamma} \pi^{-\frac{1}{2}} G_{0,3}^{3,0}\left[\frac{ab^2}{4} \big|_{0,\frac{1}{2},\gamma}\right], u \geq 0 \text{ and for } \gamma \neq \frac{\nu}{2}, \nu = 0, 1, \ldots$$

$$I_{a,b} = \Gamma\left(\frac{1}{2}\right) \Gamma(\gamma)_0 F_2\left(\ ; 1 - \gamma, \frac{1}{2}; -\frac{ab^2}{4}\right)$$

$$+ \Gamma\left(-\frac{1}{2}\right) \Gamma\left(\gamma - \frac{1}{2}\right) \left(\frac{ab^2}{4}\right)^{\frac{1}{2}} {}_0F_2\left(\ ; \frac{3}{2}, \frac{3}{2} - \gamma; -\frac{ab^2}{4}\right)$$

$$+ \Gamma(-\gamma) \Gamma\left(\frac{1}{2} - \gamma\right) \left(\frac{ab^2}{4}\right)^\gamma {}_0F_2\left(\ ; 1+\gamma, \frac{1}{2}+\gamma; -\frac{ab^2}{4}\right), u \geq 0.$$

Note 5.1 Theorem 5.1 can also be established by using statistical procedures. In this case, consider two real scalar positive random variables $x_1 > 0, x_2 > 0$, statistically independently distributed with densities $f_1(x_1)$ and $f_2(x_2)$ respectively. Due to independence, their joint density will be the product $f_1(x_1) f_2(x_2)$. Let $u = x_1 x_2$. Consider the $(s - 1)$th moment of u, that is $E[u^{s-1}]$ where $E[\cdot]$ means the expected value of $[\cdot]$. Due to independence,

$$E(u^{s-1}) = E(x_1^{s-1})E(x_2^{s-1}), \ E(u^{s-1}) = \int_0^\infty u^{s-1}g(u)du = M_g(s)$$

$$E(x_1^{s-1}) = \int_0^\infty x_1^{s-1} f_1(x_1) dx_1 = M_{f_1}(s)$$

$$E(x_2^{s-1}) = \int_0^\infty x_2^{s-1} f_2(x_2) dx_2 = M_{f_2}(s) \Rightarrow$$

$$M_g(s) = M_{f_1}(s) M_{f_2}(s)$$

where $g(u)$ is the density of u, which is to be determined. We can take the same functions $f_1(x_1)$ and $f_2(x_2)$ as in Sect. 5.1 but in this case, multiply by the respective normalizing constants to construct densities out of those f_1 and f_2. These, normalizing constants get canceled in the equation $M_g(s) = M_{f_1}(s) M_{f_2}(s)$ and hence the inverse, namely $g(u)$ will be one and the same as obtained in Sect. 5.1. The original derivation of the reaction-rate probability integral in 1984, which was reported in Mathai and Haubold (1988), was done through statistical distribution theory as described in this Note 5.1. When γ is an integer or a half-integer, then some of the poles can be of order 1 and the remaining of order 2. In this case, one can obtain computable series forms involving gamma function, psi function and logarithm, details may be seen from Mathai (1993) where the techniques of deriving the residues from poles of all types of orders are given.

5.2 Generalization of the Reaction-Rate Probability Integral

A generalization of the integral $I_{a,b}$ is the following integral:

$$I_{a,b}(\delta, \rho) = \int_0^\infty v^{\gamma-1} e^{-av^\delta - bv^{-\rho}} dv \tag{5.15}$$

for $a > 0, b > 0, \delta > 0, \rho > 0, \Re(\gamma) > 0$. For the evaluation of the integral (5.15) also, we can proceed exactly as in the case of the integral $I_{a,b}$ of Sect. 5.1. Before the integral is evaluated, let us consider some particular cases. For $\delta = 1, \rho = 1$, (5.15) is the basic Bessel integral. For $\delta = 1, \rho = 1$, the integrand in (5.15), normalized, is the inverse Gaussian density. For $\delta = 1, \rho = \frac{1}{2}$, it is the reaction-rate probability integral in nuclear reaction-rate theory. For $\delta = 1$ and general ρ, it is Krätzel integral and Krätzel transform is associated with it. Mathai (2012) has created a statistical density out of (5.15) and studied its properties. We may call (5.15) as a generalized Bessel integral. (5.15) is also known in the literature by different names such as generalized gamma integral, ultra gamma integral and super gamma integral. From the series representations, it is obvious that (5.15) does not belong to the gamma family of functions, see also Mathai (2016). For the evaluation of (5.15), consider

$$f_1(x_1) = e^{-x_1^\rho} \Rightarrow M_{f_1}(s) = \int_0^\infty x_1^{s-1} e^{-x_1^\rho} dx_1 = \frac{1}{\rho}\Gamma\left(\frac{s}{\rho}\right), \Re(s) > 0, \rho > 0 \tag{5.16}$$

$$f_2(x_2) = x_2^\gamma e^{-ax_2^\delta} \Rightarrow M_{f_2}(s) = \int_0^\infty x_2^{\gamma+s-1} e^{-ax_2^\delta} dx_2$$
$$= \frac{1}{\delta}\Gamma\left(\frac{\gamma+s}{\delta}\right) a^{-\left(\frac{\gamma+s}{\delta}\right)}, \Re(s) > -\Re(\gamma), \delta > 0, a > 0, \Re(\gamma) > 0 \tag{5.17}$$

where $M_{f_1}(s)$ and $M_{f_2}(s)$ are the Mellin transforms of f_1 and f_2 respectively, with Mellin parameter s. Let $u = x_1 x_2$ and $v = x_2$ with the joint function $f_1(x_1) f_2(x_2)$, the product. Let $g(u)$ be the function associated with u. Then,

$$g(u) = \int_0^\infty \frac{1}{v} f_1\left(\frac{u}{v}\right) f_2(v) dv = \int_0^\infty \frac{1}{v} f_1(v) f_2\left(\frac{u}{v}\right) dv \tag{5.18}$$

where the second part integral is obtained by taking $x_1 = v$ in the transformation $u = x_1 x_2$. Now, for $\Re(s) > 0, \Re(s) > -\Re(\gamma), \Re(\gamma) > 0 \Rightarrow \Re(s) > 0$,

$$M_g(s) = M_{f_1}(s) M_{f_2}(s) = \frac{1}{\rho \delta a^{\gamma/\delta}} \Gamma\left(\frac{s}{\rho}\right) \Gamma\left(\frac{\gamma+s}{\delta}\right) a^{-\frac{s}{\delta}}$$
$$= \frac{1}{\rho \delta a^{\gamma/\delta}} \Gamma\left(\frac{s}{\rho}\right) \Gamma\left(\frac{\gamma}{\delta} + \frac{s}{\delta}\right) a^{-\frac{s}{\delta}}. \tag{5.19}$$

Therefore, from the inverse Mellin transform,

$$g(u) = \frac{1}{\rho \delta a^{\gamma/\delta}} \frac{1}{2\pi i} \int_{c-i\infty}^{c+i\infty} \Gamma\left(\frac{s}{\rho}\right) \Gamma\left(\frac{\gamma}{\delta} + \frac{s}{\delta}\right) \left(a^{\frac{1}{\delta}} u\right)^{-s} ds, u = b^{\frac{1}{\rho}}, i = \sqrt{(-1)}. \tag{5.20}$$

As shown in Sect. 5.1, this (5.20) can be written in terms of a H-function, namely,

$$g(u) = \frac{1}{\rho \delta a^{\gamma/\delta}} H_{0,2}^{2,0}\left[a^{\frac{1}{\delta}} u \Big|_{(0,\frac{1}{\rho}),(\frac{\gamma}{\delta},\frac{1}{\delta})}\right], u = b^{\frac{1}{\rho}} \tag{5.21}$$

where the c in the contour is such that $c > 0, \delta > 0, \rho > 0, \Re(\gamma) > 0$. Consider the special case $\rho = \delta$. In this case, one can write the H-function in (5.21) in terms of a G-function. Replace $\frac{s}{\delta}$ by s, thereby s by δs and ds by δds. Then, we have the following:

$$g(u) = \frac{1}{\delta a^{\gamma/\delta}} \frac{1}{2\pi i} \int_{c_1-i\infty}^{c_1+i\infty} \Gamma(s) \Gamma\left(\frac{\gamma}{\delta} + s\right) (au^\delta)^{-s} ds, c_1 > 0$$
$$= \frac{1}{\delta a^{\gamma/\delta}} G_{0,2}^{2,0}\left[au^\delta \Big|_{0,\frac{\gamma}{\delta}}\right], u \geq 0, u = b^{\frac{1}{\delta}}, \rho = \delta \tag{5.22}$$

where $G(\cdot)$ is the G-function as explained in Sect. 11 or see Mathai (1993). Then, for $\frac{\gamma}{\delta} \neq \nu, \nu = 0, 1, \ldots$ the poles of the integrand in (5.22) are simple and then proceeding as in Sect. 5.1, we have the following result:

Theorem 5.2 *For the $g(u)$ defined in (vii) above, the following is the explicit computable series form when $\frac{\gamma}{\delta} \neq \nu, \nu = 0, 1, \ldots$:*

$$g(u) = \frac{1}{\delta a^{\gamma/\delta}} \left[\Gamma\left(\frac{\gamma}{\delta}\right) {}_0F_1\left(\; ; 1 - \frac{\gamma}{\delta}; au^\delta \right) \right.$$
$$\left. + \Gamma\left(-\frac{\gamma}{\delta}\right) (au^\delta)^{\frac{\gamma}{\delta}} {}_0F_1\left(\; ; 1 + \frac{\gamma}{\delta}; au^\delta \right), u = b^{\frac{1}{\delta}} \right] \quad (5.23)$$

where ${}_0F_1$ is the classical Bessel series. The details of the derivation are omitted because the derivation is parallel to that in Sect. 5.1.

5.3 An Extension Through Mathai's Pathway Idea

In a physical system the ideal behavior of a variable x, under observation, may be governed by the function $e^{-bx^{-\rho}}, b > 0, \rho > 0$. But, in a practical situation the behavior may be $e^{-bx^{-\rho}}$ or in the neighborhood of this ideal behavior. The ideal function and its neighborhoods are covered by the function

$$(1 + b(q-1)x^{-\rho})^{-\frac{1}{q-1}} \to e^{-bx^{-\rho}} \text{ for } q \geq 1. \quad (5.24)$$

For $q < 1, q - 1 = -(1-q), q < 1$ and then

$$(1 - b(1-q)x^{-\rho})^{\frac{1}{1-q}} \to e^{-bx^{-\rho}} \text{ when } 1 - b(1-q)x^{-\rho} > 0 \text{ and } q \leq 1. \quad (5.25)$$

Thus, for $-\infty < q < 1$, the binomial form in (5.25), for $1 < q < \infty$, the binomial form in (5.24), and for $q \to 1$, the exponential form $e^{-bx^{-\rho}}$, are all reached through the parameter q, and either from the binomial from in (5.24) or from the binomial form in (5.25). From these observations, Mathai (2005) proposed a rectangular matrix-variate model, known as Mathai's pathway model. If the pathway model is taken for a real scalar positive variable, then it will be either the model in (5.24), which can switch into (5.25) or vice-versa. Consider the following integral, denoted by $I_{a,b}(\delta, \rho; q)$, where the pathway idea is incorporated:

$$I_{a,b}(\delta, \rho; q) = \int_0^\infty v^{\gamma-1} e^{-av^\delta} [1 + b(q-1)v^{-\rho}]^{-\frac{1}{q-1}} dv$$
$$= \int_0^{(b(1-q))^{1/\rho}} v^{\gamma-1} e^{-av^\delta} [1 - b(1-q)v^{-\rho}]^{\frac{1}{1-q}}, q < 1, 1 - b(1-q)v^{-\rho} > 0$$
$$= I_{a,b}(\delta, \rho; 1) = I_{a,b}(\delta, \rho) \text{ for } q \to 1.$$

Again, we will evaluate the integral $I_{a,b}(\delta, \rho; q)$ by using the Mellin convolution of a product property. Let us consider the case $q > 1$ first.

Case 5.1 $q > 1$. Let $f_1(x_1) = [1 + (q-1)x_1^\rho]^{-\frac{1}{q-1}}$ and $f_2(x_2) = x_2^\gamma e^{-ax_2^\delta}$. Then,

$$M_{f_1}(s) = \int_0^\infty x_1^{s-1}[1+(q-1)x_1^\rho]^{-\frac{1}{q-1}}dx_1, b = u^\rho$$

$$= \frac{1}{\rho}[(q-1)]^{-\frac{s}{\rho}} \frac{\Gamma(\frac{s}{\rho})\Gamma(\frac{1}{q-1} - \frac{s}{\rho})}{\Gamma(\frac{1}{q-1})} \quad (5.26)$$

for $q > 1, \rho > 0, \Re(s) > 0, \Re(s) < \frac{\rho}{q-1}$ where the integral is evaluated with the help of a real scalar type-2 beta integral.

$$M_{f_2}(s) = \int_0^\infty x_2^{\gamma+s-1} e^{-ax_2^\delta} dx_2 = \frac{1}{\delta} a^{-\frac{\gamma+s}{\delta}} \Gamma\left(\frac{\gamma+s}{\delta}\right) \quad (5.27)$$

for $\Re(s) > -\Re(\gamma), \Re(\gamma) > 0, a > 0, \delta > 0$. Then, from the Mellin convolution of a product property, the Mellin transform of g, the function corresponding to $u = x_1 x_2$, is the following:

$$M_g(s) = \frac{(a^{\frac{1}{\delta}}(q-1)^{\frac{1}{\rho}})^{-s}}{\delta\rho a^{\gamma/\delta}\Gamma(\frac{1}{q-1})} \Gamma\left(\frac{s}{\rho}\right) \Gamma\left(\frac{\gamma+s}{\delta}\right) \Gamma\left(\frac{1}{q-1} - \frac{s}{\rho}\right) \quad (5.28)$$

for $\delta > 0, \rho > 0, a > 0, q > 1, 0 < \Re(s) < \frac{\rho}{q-1}, b = u^\rho$. Hence, from the inverse Mellin transform, the function $g(u)$ is available as the following, where $c^{-1} = \rho\delta a^{\gamma/\delta}\Gamma\left(\frac{1}{q-1}\right)$:

$$g(u) = c\frac{1}{2\pi i}\int_{c-i\infty}^{c+i\infty} \Gamma\left(\frac{s}{\rho}\right)\Gamma\left(\frac{\gamma+s}{\delta}\right)\Gamma\left(\frac{1}{q-1} - \frac{s}{\rho}\right)[a^{\frac{1}{\delta}}(q-1)^{\frac{1}{\rho}}u]^{-s}ds$$

$$= cH_{1,2}^{2,1}\left[a^{\frac{1}{\delta}}(q-1)^{\frac{1}{\rho}}u \Big|_{(0,\frac{1}{\delta}),(\frac{\gamma}{\delta},\frac{1}{\delta})}^{(1-\frac{1}{q-1},\frac{1}{\rho})}\right], u \geq 0, u = b^{\frac{1}{\rho}}. \quad (5.29)$$

But, for $\rho = \delta$, we may replace $\frac{s}{\delta}$ by s thereby s by δs and ds by δds. Then, $g(u)$ of (5.29) can be written in terms of a G-function as the following:

$$g(u) = \frac{1}{\delta a^{\gamma/\delta}\Gamma(\frac{1}{q-1})} G_{1,2}^{2,1}\left[ab(q-1)\Big|_{0,\frac{\gamma}{\delta}}^{1-\frac{1}{q-1}}\right], u \geq 0, \rho = \delta, u = b^{\frac{1}{\rho}}, q > 1. \quad (5.30)$$

That is, $g(u)$ of (5.30) is given by the following, for $c_1^{-1} = \delta a^{\gamma/\delta}\Gamma(\frac{1}{q-1}), q > 1$:

$$g(u) = c_1 \frac{1}{2\pi i}\int_{c-i\infty}^{c+i\infty} \Gamma(s)\Gamma\left(\frac{\gamma}{\delta}+s\right)\Gamma\left(\frac{1}{q-1} - s\right)[ab(q-1)]^{-s}ds \quad (5.31)$$

5.3 An Extension Through Mathai's Pathway Idea

where the c in the contour is such that $0 < c < \frac{\rho}{q-1}$. This G-function in (5.31) can be evaluated as the sum of the residues at the poles of $\Gamma(s)\Gamma(\frac{\gamma}{\delta}+s)$. The continuation part gives a divergent series. The poles of $\Gamma(s)\Gamma(\frac{\gamma}{\delta}+s)$ are simple if $\frac{\gamma}{\delta} \neq \nu, \nu = 0, 1, \ldots$. Hence, in this special case we will write down the computable series form for the function $g(u)$. The steps are parallel to those used in Sects. 5.1 and 5.2 and hence the final result will be stated here as a theorem.

Theorem 5.3 *For $\delta = \rho, q > 1, \frac{\gamma}{\delta} \neq \nu, \nu = 0, 1, \ldots, g(u)$ of (5.31) above is the following, where c_1 is given in (5.31) above and $u = b^{\frac{1}{\rho}}, \rho = \delta, q > 1$:*

$$g(u) = c_1 \left[\Gamma\left(\frac{1}{q-1}\right) \Gamma\left(\frac{\gamma}{\delta}\right) {}_1F_1\left(\frac{1}{q-1}; 1 - \frac{\gamma}{\delta}; ab(q-1)\right) \right.$$
$$\left. + \Gamma\left(\frac{1}{q-1} + \frac{\gamma}{\delta}\right) \Gamma\left(-\frac{\gamma}{\delta}\right) (ab(q-1))^{\frac{\gamma}{\delta}} {}_1F_1\left(\frac{1}{q-1} + \frac{\gamma}{\delta}; 1 + \frac{\gamma}{\delta}; ab(q-1)\right) \right]. \tag{5.32}$$

Observe that the confluent hypergeometric series ${}_1F_1$ is convergent for all values of $ab(q-1)$.

Case 5.2 $q < 1$. Let $f_1(x_1) = [1 - (1-q)x_1^{\rho}]^{\frac{1}{1-q}}, q < 1$ and $f_2(x_2) = x_2^{\gamma} e^{-ax_2^{\delta}}$. Then, their Mellin transforms are the following, where the computational steps are parallel to those in the case $q > 1$ and hence the details are not given here:

$$M_{f_1}(s) = \frac{1}{\rho} \Gamma\left(\frac{s}{\rho}\right) (1-q)^{-\frac{s}{\rho}} \frac{\Gamma(\frac{1}{1-q} + 1)}{\Gamma(\frac{1}{1-q} + 1 + \frac{s}{\rho})}$$

for $\Re(s) > 0, \rho > 0, q < 1$.

$$M_{f_2}(s) = \frac{1}{\delta} \Gamma\left(\frac{\gamma + s}{\delta}\right) a^{-\frac{\gamma+s}{\delta}}$$

for $\Re(s) > -\Re(\gamma) > 0, \Re(\gamma) > 0, a > 0, \delta > 0$. Therefore, the Mellin transform of $u = x_1 x_2$, when the joint function is $f_1(x_1) f_2(x_2)$, is given by the following:

$$M_g(s) = \frac{\Gamma(\frac{1}{1-q}+1)}{\rho \delta a^{\gamma/\delta}} \frac{\Gamma(\frac{s}{\rho})\Gamma(\frac{\gamma+s}{\delta})}{\Gamma(1 + \frac{1}{1-q} + \frac{s}{\rho})} (a^{\frac{1}{\delta}}(1-q)^{\frac{1}{\rho}})^{-s}$$

for $\delta > 0, \rho > 0, q < 1, \Re(\gamma) > 0, a > 0, \Re(s) > 0, u^{\rho} = b$. Therefore, from the inverse Mellin transform, the function $g(u)$ for $u^{\rho} = b$ is the following, for $c_2 = \frac{\Gamma(1+\frac{1}{1-q})}{\delta \rho a^{\gamma/\delta}}$:

$$g(u) = c_2 \frac{1}{2\pi i} \int_{c-i\infty}^{c+i\infty} \frac{\Gamma(\frac{s}{\rho})\Gamma(\frac{\gamma+s}{\delta})}{\Gamma(1+\frac{1}{1-q}+\frac{s}{\rho})} (a^{\frac{1}{\delta}}(b(1-q))^{\frac{1}{\rho}})^{-s} ds$$

$$= c_2 H_{0,2}^{2,0}\left[a^{\frac{1}{\delta}}(b(1-q))^{\frac{1}{\rho}} \Big|_{(0,\frac{1}{\rho}),(\frac{\gamma}{\delta},\frac{1}{\delta})}^{(1+\frac{1}{1-q},\frac{1}{\rho})} \right], q < 1 \quad (5.33)$$

for $a > 0, \delta > 0, \rho > 0, \Re(\gamma) > 0, b > 0, q < 1$. As in the case for $q > 1$, here also we can represent $g(u)$ in terms of a G-function for the case $\rho = \delta$. Let $\rho = \delta$. Now, replace $\frac{s}{\delta}$ by s thereby s by δs and ds by δds. Then, c_2 of (5.33) changes to c_2' where c_2' is the same c_2 with ρ removed from the denominator because this is canceled from the δ coming from δds. That is,

$$g(u) = c_2' \frac{1}{2\pi i} \int_{c-i\infty}^{c+i\infty} \frac{\Gamma(s)\Gamma(\frac{\gamma}{\delta}+s)}{\Gamma(1+\frac{1}{1-q}+s)} [ab(1-q)]^{-s} ds$$

$$= c_2' G_{0,2}^{2,0}\left[ab(1-q) \Big|_{0,\frac{\gamma}{\delta}}^{1+\frac{1}{1-q}} \right], \rho = \delta, q < 1 \quad (5.34)$$

where the c in the contour is any real number > 0. Note that for $\frac{\gamma}{\delta} \neq \nu, \nu = 0, 1, \ldots$ the poles of the integrand in (5.34) are simple. Hence, in this case, proceeding as in the case of $q > 1$, we have computable series form, which will be stated as a theorem.

Theorem 5.4 *For $q < 1, \rho = \delta, \frac{\gamma}{\delta} \neq \nu, \nu = 0, 1, \ldots, g(u)$ in (5.34) is given by the following series forms:*

$$g(u) = \frac{\Gamma(1+\frac{1}{1-q})}{\delta a^{\gamma/\delta}} \left[\frac{\Gamma(\frac{\gamma}{\delta})}{\Gamma(1+\frac{1}{1-q})} {}_1F_1\left(-\frac{1}{1-q}; 1-\frac{\gamma}{\delta}; -ab(1-q) \right) \right.$$
$$\left. + \frac{\Gamma(-\frac{\gamma}{\delta})}{\Gamma(1+\frac{1}{1-q}-\frac{\gamma}{\delta})} [ab(1-q)]^{\frac{\gamma}{\delta}} {}_1F_1\left(\frac{\gamma}{\delta} - \frac{1}{1-q}; 1+\frac{\gamma}{\delta}; -ab(1-q) \right) \right]$$
(5.35)

for $a > 0, \delta > 0, a > 0, b > 0, q < 1, \Re(\gamma) > 0, u^\rho = b, \rho = \delta$.

Note 5.2 Note that

$$\lim_{q_1 \to 1} [1 + a(q_1-1)x_2^\delta]^{-\frac{1}{q_1-1}} = e^{-ax_2^\delta}.$$

Hence, one may consider an extended integral $I_{a,b}(\delta, \rho)$ as the following, denoted by $I_{a,b}(\delta, \rho; q_1, q_2)$, where

$$I_{a,b}(\delta, \rho; q_1, q_2) = \int_0^\infty v^{\gamma-1}[1+a(q_1-1)v^\delta]^{-\frac{1}{q_1-1}} [1+b(q_2-1)v^{-\rho}]^{-\frac{1}{q_2-1}} dv$$

for $q_1 > 1$ or $q_1 < 1$ or $q_1 \to 1$ and then independently $q_2 > 1, q_2 < 1, q_2 \to 1$. Thus, there are nine situations in the above integral. Such an integral called "a versatile integral" was evaluated in Mathai and Haubold (2019). The above versatile integral can also be handled by using the procedures in Sects. 5.1, 5.2, 5.3. Hence, this will not be discussed in detail here.

5.4 The Pathway Extended Reaction-Rate Probability Integral

The following are three different extended forms of the basic reaction-rate probability integral in the real scalar positive variable case:

$$I_{a,b}(q_1) = \int_0^\infty v^{\gamma-1}[1+a(q_1-1)v]^{-\frac{1}{q_1-1}} e^{-bv^{-\frac{1}{2}}} dv, \; q_1 > 1 \quad (5.36)$$

$$I_{a,b}(q_2) = \int_0^\infty v^{\gamma-1} e^{-av}[1+b(q_2-1)v^{-\frac{1}{2}}]^{-\frac{1}{q_2-1}} dv, \; q_2 > 1 \quad (5.37)$$

$$I_{a,b}(q_1,q_2) = \int_0^\infty v^{\gamma-1}[1+a(q_1-1)v]^{-\frac{1}{q_1-1}}$$
$$\times [1+b(q_2-1)v^{-\frac{1}{2}}]^{-\frac{1}{q_2-1}} dv, \; q_j > 1, j=1,2. \quad (5.38)$$

Observe that in the first two integrals, (5.36), (5.37), there are three situations each, namely $q_j > 1, q_j < 1, q_j \to 1$. In the third case (5.28) there are nine situations, and each situation is an extended form of the reaction-rate integral except the original basic integral. The techniques used in Sects. 5.1, 5.2, 5.3s can be applied to derive the H-function format or G-function format or series forms for the left side in each of (5.36)-(5.38). For the sake of illustration of the techniques, we will evaluate one of the above integrals.

Case 5.3 Evaluation of the first integral, $q_1 > 1$

The integral to be evaluated is the following:

$$I_{a,b}(q_1) = \int_0^\infty v^{\gamma-1}[1+a(q_1-1)v]^{-\frac{1}{q_1-1}} e^{-bv^{-\frac{1}{2}}} dv, \; q_1 > 1.$$

Let $f_1(x_1) = e^{-x_1^{\frac{1}{2}}}$ and $f_2(x_2) = x_2^\gamma[1+a(q_1-1)x_2]^{-\frac{1}{q_1-1}}$. Then, $f_1(\frac{u}{v}) = e^{-(\frac{u}{v})^{\frac{1}{2}}} = e^{-bv^{-\frac{1}{2}}}, b = u^{\frac{1}{2}} \Rightarrow u = b^2$.

$$M_{f_1}(s) = \int_0^\infty x_1^{s-1} e^{-x_1^{\frac{1}{2}}} dx_1$$
$$= 2 \int_0^\infty y^{2s-1} e^{-y} dy$$
$$= 2\Gamma(2s) = \pi^{-\frac{1}{2}} 2^{2s} \Gamma(s) \Gamma\left(s + \frac{1}{2}\right), \tag{5.39}$$

for $\Re(s) > 0$. Observe that $\Gamma(2s)$ is expanded by using the duplication formula for gamma functions, which was already illustrated before.

$$M_{f_2}(s) = \int_0^\infty x_2^{\gamma+s-1} [1 + a(q_1 - 1)x_2]^{-\frac{1}{q_1-1}} dx_2$$
$$= [a(q_1 - 1)]^{-(\gamma+s)} \frac{\Gamma(\gamma+s)\Gamma(\frac{1}{q_1-1} - (\gamma+s))}{\Gamma(\frac{1}{q_1-1})}$$

for $\Re(s) > -\Re(\gamma), \Re(s) < \frac{1}{q_1-1} - \Re(\gamma), q_1 > 1, a > 0, \Re(\gamma) > 0$. Then, from the Mellin convolution property, the Mellin transform of g is the following:

$$M_g(s) = \frac{[a(q_1-1)]^{-s} 4^s}{[a(q_1-1)]^\gamma \pi^{\frac{1}{2}} \Gamma(\frac{1}{q_1-1})}$$
$$\times \Gamma(s) \Gamma\left(s + \frac{1}{2}\right) \Gamma(\gamma+s) \Gamma\left(\frac{1}{q_1-1} - \gamma - s\right)$$

for $q_1 > 1, a > 0, \Re(\gamma) > 0, 0 < \Re(s) < \frac{1}{q_1-1}$. Then, from the inverse Mellin transform $g(u)$, which is the integral to be evaluated, is the following:

$$g(u) = c_1 \frac{1}{2\pi i} \int_{c-i\infty}^{c+i\infty} \Gamma(s) \Gamma\left(s + \frac{1}{2}\right) \Gamma(\gamma+s)$$
$$\times \Gamma\left(\frac{1}{q_1-1} - \gamma - s\right) \left[\frac{a(q_1-1)u}{4}\right]^{-s} ds$$
$$= c_1 \, G_{1,3}^{3,1} \left[\frac{a(q_1-1)u}{4} \bigg|_{0,\frac{1}{2},\gamma}^{1+\gamma-\frac{1}{q_1-1}}\right] \tag{5.40}$$
$$c_1^{-1} = \pi^{\frac{1}{2}} [a(q_1-1)]^\gamma \Gamma\left(\frac{1}{q_1-1}\right), q_1 > 1, a > 0, u = b^2, b > 0.$$

Observe that one can evaluate (5.40) as the sum of the residues at the poles of $\Gamma(s)\Gamma(s + \frac{1}{2})\Gamma(\gamma+s)$ because the continuation part is divergent. The poles are simple when γ is not an integer or half-integer. We will obtain the series form for this case. The method is the same as the ones used in Sects. 5.1–5.3 Hence, the final result will be written as a theorem.

5.4 The Pathway Extended Reaction-Rate Probability Integral

Theorem 5.5 *For $\gamma \neq \frac{\nu}{2}, \nu = 0, 1, \ldots, g(u)$ of (5.40) is the following for $u = b^2, a > 0, b > 0, q_1 > 1$:*

$$g(u) = \int_0^\infty v^{\gamma-1}[1 + a(q_1 - 1)v]^{-\frac{1}{q_1-1}} e^{-bv^{-\frac{1}{2}}} dv$$

$$= \frac{1}{\pi^{\frac{1}{2}}[a(q_1-1)]^\gamma \Gamma(\frac{1}{q_1-1})} \left[\Gamma\left(\frac{1}{2}\right) \Gamma(\gamma) \Gamma\left(\frac{1}{q_1-1} - \gamma\right) \right.$$

$$\times {}_1F_2\left(\frac{1}{q_1-1} - \gamma; \frac{1}{2}, 1-\gamma; -\frac{a(q_1-1)b^2}{4}\right)$$

$$+ \Gamma\left(-\frac{1}{2}\right) \Gamma\left(\gamma - \frac{1}{2}\right) \Gamma\left(\frac{1}{q_1-1} - \gamma + \frac{1}{2}\right) \left[\frac{a(q_1-1)b^2}{4}\right]^{\frac{1}{2}}$$

$$\times {}_1F_2\left(\frac{1}{q_1-1} - \gamma + \frac{1}{2}; \frac{3}{2}, \frac{3}{2} - \gamma; -\frac{a(q_1-1)b^2}{4}\right)$$

$$+ \Gamma(-\gamma) \Gamma\left(\frac{1}{2} - \gamma\right) \Gamma\left(\frac{1}{q_1-1}\right) \left[\frac{s(q_1-1)b^2}{4}\right]^\gamma$$

$$\left. \times {}_1F_2\left(\frac{1}{q_1-1}; 1+\gamma, \frac{1}{2}+\gamma; -\frac{a(q_1-1)b^2}{4}\right) \right] \quad (5.41)$$

Case 5.4 First integral, $q_1 < 1$

Consider

$$I_{a,b}(q_1) = \int_0^{\frac{1}{a(1-q_1)}} v^{\gamma-1}[1 - a(1-q_1)v]^{\frac{1}{1-q_1}} e^{-bv^{-\frac{1}{2}}} dv$$

for $q_1 < 1, a > 0, b > 0, 1 - a(1-q_1)v > 0$. We use the same technique as before. Here $f_1(x_1)$ remains the same with $M_{f_1}(s) = \pi^{-\frac{1}{2}} \Gamma(s) \Gamma(s + \frac{1}{2}) 4^s, \Re(s) > 0$. But, $f_2(x_2)$ will be different;

$$f_2(x_2) = x_2^\gamma[1 - a(1-q_1)x_1]^{\frac{1}{1-q_1}} \Rightarrow$$

$$M_{f_2}(s) = [a(1-q_1)]^{-(\gamma+s)} \frac{\Gamma(\gamma+s)\Gamma(1+\frac{1}{1-q_1})}{\Gamma(1+\frac{1}{1-q_1}+s)}$$

for $\Re(s) > -\Re(\gamma), \Re(\gamma) > 0$. From the Mellin convolution of the product property, the Mellin transform of $g(u)$ is the following:

$$M_g(s) = \frac{\Gamma(1+\frac{1}{1-q_1})}{\pi^{\frac{1}{2}}[a(1-q_1)]^\gamma} \left[\frac{a(1-q_1)}{4}\right]^{-s}$$

$$\times \frac{\Gamma(s)\Gamma(s+\frac{1}{2})\Gamma(\gamma+s)}{\Gamma(1+\frac{1}{1-q_1}+s)}$$

for $a > 0, q_1 < 1, \Re(s) > 0, \Re(\gamma) > 0$. From the inverse Mellin transform, we have $g(u)$ as the following:

$$g(u) = c_1 \frac{1}{2\pi i} \int_{c-i\infty}^{c+i\infty} \frac{\Gamma(s)\Gamma(s+\frac{1}{2})\Gamma(\gamma+s)}{\Gamma(1+\frac{1}{1-q_1}+s)} \left[\frac{a(1-q_1)b^2}{4}\right]^{-s} ds$$

$$= c_1 G_{1,3}^{3,0} \left[\frac{a(1-q_1)b^2}{4} \Big|_{0,\frac{1}{2},\gamma}^{1+\frac{1}{1-q_1}}\right], u = b^2$$

$$c_1 = \frac{\Gamma(1+\frac{1}{1-q_1})}{\pi^{\frac{1}{2}}[a(1-q_1)]^{\gamma}} \tag{5.42}$$

where c in the contour is any real number > 0, $a > 0, b > 0, \Re(\gamma) > 0, q_1 < 1$. Note that, as in the previous case, we can obtain a series from for $g(u)$ when $\gamma \neq \frac{\nu}{2}, \nu = 0, 1, \ldots$ in which case the poles of $\Gamma(s)\Gamma(s+\frac{1}{2})\Gamma(\gamma)$ are simple. The result will be stated as a theorem.

Theorem 5.6 *For $\gamma \neq \frac{\nu}{2}, \nu = 0, 1, \ldots$, the $g(u)$ for case (2) in (5.42) above, is the following, where c_1 is given in (5.42) above:*

$$g(u) = c_1 \left[\frac{\Gamma(\frac{1}{2})\Gamma(\gamma)}{\Gamma(1+\frac{1}{1-q_1})} {}_1F_2\left(-\frac{1}{1-q_1}; \frac{1}{2}, 1-\gamma; \frac{a(1-q_1)b^2}{4}\right) \right.$$

$$+ \frac{\Gamma(-\frac{1}{2})\Gamma(\gamma-\frac{1}{2})}{\Gamma(\frac{1}{2}+\frac{1}{1-q_1})} \left[\frac{a(1-q_1)b^2}{4}\right]^{\frac{1}{2}}$$

$$\times {}_1F_2\left(\frac{1}{2}-\frac{1}{1-q_1}; \frac{3}{2}, \frac{3}{2}-\gamma; \frac{a(1-q_1)b^2}{4}\right)$$

$$+ \frac{\Gamma(-\gamma)\Gamma(\frac{1}{2}-\gamma)}{\Gamma(1+\frac{1}{1-q_1}-\gamma)} \left[\frac{a(1-q_1)b^2}{4}\right]^{\gamma}$$

$$\left. \times {}_1F_2\left(\gamma-\frac{1}{1-q_1}; 1+\gamma, \frac{1}{2}+\gamma; \frac{a(1-q_1)b^2}{4}\right) \right], q_1 < 1 \tag{5.43}$$

Note 5.3 If the integrals considered in Sects. 5.1–5.3 are extended to the whole real line, then it can be achieved by replacing v in the integral by $|v|$, the absolute value of v. For example, the generalized reaction-rate integral will then be of the following form:

$$\int_{-\infty}^{\infty} |x|^{\gamma-1} e^{-|x|^{\delta} - b|x|^{-\rho}} dx \tag{5.44}$$

for $a > 0, b > 0, \delta > 0, \rho > 0, \Re(\gamma) > 0$. Corresponding changes may be made in all other integrals such as $I_{a,b}, I_{a,b}(q_1), I_{a,b}(q_2), I_{a,b}(q_1, q_2)$. Mellin convolution of the product of Maxwell-Boltzmann and Raleigh densities and their extended forms to the whole line are covered in (5.44).

5.5 Generalized Reaction-Rate Probability Integral in the Real Multivariate Case

There may not be any corresponding physics or reaction-rate probability integral or Krätzel integral yet in what we are going to discuss in this section. These results may be motivating factors for developing corresponding physics or communication theory or engineering problems later on. Let X be a $p \times 1$ real vector with $X' = [x_1, \ldots, x_p]$, where a prime denotes the transpose and the x_j's are distinct (functionally independent) real scalar variables, $-\infty < x_j < \infty, j = 1, \ldots, p$. Then, $y = X'X = x_1^2 + \cdots + x_p^2$ is an isotropic quantity, in the sense that y remains invariant under the rotation of the axes of coordinates or under orthonormal transformations, that is, if $Z = AX$, $AA' = I_p$, $A'A = I_p$, then $Z'Z = X'X$. Such isotropic variables appear in different disciplines. $X'X$ is associated with isotropic random points in p-dimensional Euclidean space in geometrical probability problems, see Mathai (1999). Also, $X'X$ is associated with spherically symmetric distributions in statistical distribution theory and related areas. Let us consider the integral, denoted by $\mathbf{I}_{a,b}(\delta, \rho)$ with I in bold,

$$\mathbf{I}_{a,b}(\delta, \rho) = \int_X (X'X)^\gamma e^{-a(X'X)^\delta - b(X'X)^{-\rho}} dX \qquad (5.45)$$

for $a > 0, b > 0, \delta > 0, \rho > 0$.

The following general notations will be used in this and coming sections. Real scalar variables, whether they are mathematical variables or random variables, will be denoted by lower-case letters such as x, y. Vector (a $p \times 1$ or $1 \times p$ matrix)/matrix variables will be denoted by capital letters such as X, Y, whether the variables are mathematical or random. Scalar constants will be denoted by a, b etc. and vector/matrix constants by A, B etc. Variables in the complex domain will be written with a tilde such as $\tilde{x}, \tilde{y}, \tilde{X}, \tilde{Y}$. No tilde will be used on constants. The determinant of a $p \times p$ matrix A will be written as $|A|$ or as $\det(A)$. When A is in the complex domain, then $|A| = a + ib, i = \sqrt{(-1)}, a, b$ are real scalars. Then, the absolute value of the determinant is written as $|\det(A)| = \sqrt{a^2 + b^2}$. Also, A^* will denote the conjugate transpose of A, denoting the conjugate by A^c and transpose by A'. For the $p \times q$ matrix $X = (x_{ij})$, where the elements x_{ij}'s are distinct real scalar variables, the wedge product of differentials will be denoted as $dX = \wedge_{i=1}^p \wedge_{j=1}^q dx_{ij}$. If the $p \times p$ matrix $Y = (y_{ij}) = Y'$ (real symmetric), then $dY = \wedge_{i \geq j} dy_{ij} = \wedge_{i \leq j} dy_{ij}$. For two real scalar variables x and y with differentials dx and dy, the wedge ($\wedge$) product of the differentials is defined as $dx \wedge dy = -dy \wedge dx \Rightarrow dx \wedge dx = 0, dy \wedge dy = 0$. When the $p \times q$ matrix $\tilde{X} = (\tilde{x}_{jk})$ is in the complex domain, then one can write $\tilde{X} = X_1 + iX_2, i = \sqrt{(-1)}, X_1, X_2$ are $p \times q$ real matrices. Then, $d\tilde{X}$ will be defined as $d\tilde{X} = dX_1 \wedge dX_2$. In the coming sections, we will be considering only real-valued scalar functions, the argument may be scalar/vector/matrix in the real or complex domain. A statistical density will be defined as a real-valued scalar function $f(X)$ of X such that $f(X) \geq 0$ in the domain of X and $\int_X f(X) dX = 1$ where X

may be scalar/vector/matrix or a sequence of matrices in the real or complex domain but $f(X)$ has to be a real-valued scalar function.

In the following discussions, a result on Jacobian of matrix transformation is going to be frequently used. This will be stated as a lemma here without proof. For the proof and for other related materials, see Mathai (1997).

Lemma 5.1 *Let the $p \times q$, $p \leq q$ matrix $X = (x_{jk})$ of rank p in the real domain. Let $Y = XX'$. Then, $Y > O$ (real positive definite). Then, going through a transformation involving a lower triangular matrix with positive diagonal elements and a unique semi-orthonormal matrix and after integrating out the differential element corresponding to the semi-orthonormal matrix, the following connection is obtained between the differential elements dX and dY:*

$$dX = \frac{\pi^{\frac{pq}{2}}}{\Gamma_p(\frac{q}{2})} |Y|^{\frac{q}{2} - \frac{p+1}{2}} dY$$

where, for example, $\Gamma_p(\alpha)$ is the real matrix-variate gamma function defined by the following:

$$\Gamma_p(\alpha) = \pi^{\frac{p(p-1)}{4}} \Gamma(\alpha) \Gamma\left(\alpha - \frac{1}{2}\right) \cdots \Gamma\left(\alpha - \frac{p-1}{2}\right), \Re(\alpha) > \frac{p-1}{2}$$

$$= \int_{Z>O} |Z|^{\alpha - \frac{p+1}{2}} e^{-\text{tr}(Z)} dZ, \Re(\alpha) > \frac{p-1}{2}$$

where $\text{tr}(\cdot)$ means the trace of $(\cdot)$ and Z is a $p \times p$ real positive definite matrix. Let $\tilde{X}$ be a $p \times q$, $p \leq q$ matrix of rank p in the complex domain with distinct complex scalar variables as elements. Let $\tilde{Y} = \tilde{X}\tilde{X}^* > O$ (Hermitian positive definite), where $\tilde{X}^*$ is the conjugate transpose of $\tilde{X}$. Then, going through a transformation involving a lower triangular matrix with real and positive diagonal elements and a unique semi-unitary matrix and then integrating out the differential element corresponding to the semi-unitary matrix, one has the following connection:

$$d\tilde{X} = \frac{\pi^{pq}}{\tilde{\Gamma}_p(q)} |\det(\tilde{Y})|^{q-p} d\tilde{Y}$$

where, for example, the complex matrix-variate gamma function is defined as the following:

$$\tilde{\Gamma}_p(\alpha) = \pi^{\frac{p(p-1)}{2}} \Gamma(\alpha) \Gamma(\alpha - 1) \cdots \Gamma(\alpha - p + 1), \Re(\alpha) > p - 1$$

$$= \int_{\tilde{Z}>O} |\det(\tilde{Z})|^{\alpha - p} e^{-\text{tr}(\tilde{Z})} d\tilde{Z}$$

where $\tilde{Z}$ is a $p \times p$ Hermitian positive definite matrix.

Let us continue the evaluation of our integral in (5.45). For a $1 \times p$ matrix we can apply Lemma 5.1, to $y = X'X$ considering X' be that $1 \times p$ matrix. Then,

$$dX = \frac{\pi^{\frac{p}{2}}}{\Gamma(\frac{p}{2})} y^{\frac{p}{2}-1} dy. \tag{5.46}$$

Then,

$$\mathbf{I}_{a,b}(\delta, \rho) = \frac{\pi^{\frac{p}{2}}}{\Gamma(\frac{p}{2})} \int_0^\infty y^{\gamma+\frac{p}{2}-1} e^{-ay^\delta - by^{-\rho}} dy. \tag{5.47}$$

Now, comparing with our integral in earlier sections, the only change is γ there is replaced by $\gamma + \frac{p}{2}$. Hence, the explicit evaluation is available from the results in earlier sections and hence further discussion is omitted. Now, for X a $p \times 1$ vector in the real domain, consider the following extended integrals, denoted by I_1, I_2, I_3 for $q_1 > 1, q_2 > 1$:

$$I_1 = \int_X [X'X]^\gamma [1 + a(q_1-1)(X'X)^\delta]^{-\frac{1}{q_1-1}} e^{-b(X'X)^{-\rho}} dX \tag{5.48}$$

$$I_2 = \int_X [X'X]^\gamma e^{-a(X'X)^\delta} [1 + b(q_2-1)(X'X)^{-\rho}]^{-\frac{1}{q_2-1}} dX \tag{5.49}$$

$$I_3 = \int_X [X'X]^\gamma [1 + a(q_1-1)(X'X)^\delta]^{-\frac{1}{q_1-1}}$$
$$\times [1 + b(q_2-1)(X'X)^{-\rho}]^{-\frac{1}{q_2-1}} dX. \tag{5.50}$$

Consider again, $y = X'X$ and from (5.46) above, the differential element is available. Then, in (5.48)–(5.50) we obtain a multiplicative factor $\frac{\pi^{\frac{p}{2}}}{\Gamma(\frac{p}{2})}$, the parameter γ becomes $\gamma + \frac{p}{2} - 1$ and $X'X$ is replaced by y. Then, (5.48)–(5.50) become the corresponding integrals of earlier sections and hence further discussion is omitted.

5.6 Real Matrix-Variate Case

Let $X = (x_{jk})$ be a $p \times q$, $p \leq q$ matrix of rank p in the real domain with pq distinct real scalar variables as the elements x_{jk}'s. Then,

$$\text{tr}(XX') = \sum_{j=1}^p \sum_{k=1}^q x_{jk}^2 \tag{5.51}$$

because for any real matrix $A = (a_{jk})$, $\text{tr}(AA') = \text{tr}(A'A) =$ the sum of squares of all elements in A. Now, think of (5.46) as coming from a $1 \times pq$ vector U so that UU' is the sum of squares in (5.46). Hence, we may use the result from the previous section on real vector-variate case. Here, one has pq variables instead of p variables in the previous section. Now,

$$\mathrm{d}X = \frac{\pi^{\frac{pq}{2}}}{\Gamma(\frac{pq}{2})} y^{\frac{pq}{2}-1} \mathrm{d}y, \ y = \text{tr}(XX'). \tag{5.52}$$

For a $p \times q$, $p \leq q$ matrix of rank p in the real domain, let

$$M_{a,b}(\delta, \rho) = \int_X [\text{tr}(XX')]^\gamma e^{-a[\text{tr}(XX')]^\delta - b[\text{tr}(XX')]^{-\rho}} \mathrm{d}X \tag{5.53}$$

$$= \frac{\pi^{\frac{pq}{2}}}{\Gamma(\frac{pq}{2})} \int_0^\infty y^{\gamma + \frac{pq}{2} - 1} e^{-ay^\delta - by^{-\rho}} \mathrm{d}y \tag{5.54}$$

for $y = \text{tr}(XX')$, $a > 0$, $b > 0$, $\delta > 0$, $\rho > 0$. Now, the integral in (5.54) is available from the corresponding real scalar variable case with γ replaced by $\gamma + \frac{pq}{2}$ and hence further discussion is omitted. If there is a multiplicative factor in terms of a determinant such as $|XX'|^\gamma$ and trace factor $[\text{tr}(XX')]^\eta$, then the integral in (5.53) becomes quite general. Can we evaluate such an extended Bessel integral or reaction-rate probability integral. This will be explored next.

5.7 Most General Real Matrix-Variate Case

Consider again a $p \times q$, $p \leq q$ real matrix of rank p with pq distinct real scalar variables as elements. For $\delta > 0$, $\rho > 0$, $a > 0$, $b > 0$, $\Re(\gamma) > \frac{p-1}{2}$, $\Re(\eta) > 0$, consider the integral

$$M_{a,b}(\delta, \rho; \gamma, \eta) = \int_X |XX'|^\gamma [\text{tr}(XX')]^\eta$$
$$\times e^{-a[\text{tr}(XX')]^\delta - b[\text{tr}(XX')]^{-\rho}} \mathrm{d}X. \tag{5.55}$$

Note that, even though X has pq real scalar variables as elements, $Y = XX'$ is symmetric and positive definite with $p(p+1)/2$ distinct real scalar variables in Y. Let $k = p(p+1)/2$. One can go from the differential element $\mathrm{d}X$ to the differential element $\mathrm{d}T$ where T is a lower triangular matrix with positive diagonal elements by going from X to $Y = XX'$ and then from Y to T or one can go directly from X to T by using another result from Mathai (1997), which will be stated here without proof.

Lemma 5.2 *Let $X = (x_{jk})$ be a $p \times q$, $p \leq q$ matrix of rank p in the real domain with pq distinct real scalar variables as elements x_{jk}'s. Let T be a lower trian-*

5.7 Most General Real Matrix-Variate Case

gular matrix with positive diagonal elements t_{jj}'s. Let U be a $p \times q$ unique semi-orthonormal matrix, $UU' = I_p$. Let $X = TU$. A unique choice of U can be made by putting the condition that the first nonzero element in every row is positive. Then,

$$X = TU, UU' = I_p \Rightarrow dX = \frac{\pi^{\frac{pq}{2}}}{\Gamma_p(\frac{q}{2})} \left\{ \prod_{j=1}^{p} t_{jj}^{q-j} \right\} dT.$$

In the corresponding complex case, $\tilde{X}$ is a $p \times q$, $p \leq q$ matrix of rank p in the complex domain, $\tilde{T}$ is a lower triangular matrix in the complex domain with real and positive diagonal elements t_{jj}'s and $\tilde{U}$ is a unique semi-unitary matrix $\tilde{U}\tilde{U}^* = I_p$. Then,

$$\tilde{X} = \tilde{T}\tilde{U}, \tilde{U}\tilde{U}^* = I_p \Rightarrow d\tilde{X} = \frac{\pi^{pq}}{\tilde{\Gamma}_p(q)} \left\{ \prod_{j=1}^{p} t_{jj}^{2(q-j)+1} \right\} d\tilde{T}.$$

Now, continuing with the evaluation of our integral in (5.55), observe that

$$|XX'| = \prod_{j=1}^{p} t_{jj}^2, \; \text{tr}(XX') = \text{tr}(TT') = \sum_{j \geq r} t_{jr}^2 \quad (5.56)$$

in the real case, and in the complex case, $\tilde{t}_{jr} = t_{jr1} + i t_{jr2}, i = \sqrt{(-1)}, t_{jr1}, t_{jr2}$ are real scalar variables and then

$$|\tilde{t}_{jr}|^2 = t_{jr1}^2 + t_{jr2}^2. \quad (5.57)$$

Let the total number of t_{jr}'s be k. Then, in the real case $k = p(p+1)/2$ and in the complex case $k = p^2$. The sum of squares of all the real elements in T or $\tilde{T}$ can be made into a single scalar quantity r^2 where r is the polar radius, through a general polar coordinate transformation. This transformation and the associated Jacobian will be stated as a lemma without proof. For the proof and other details, see Mathai (1997).

Lemma 5.3 *Let $x_1, \ldots, x_k$ be real scalar variables, $-\infty < x_j < \infty$, $j = 1, \ldots, k$. Consider the following transformation to r and θ_j's:*

$$x_1 = r \sin \theta_1$$
$$x_2 = r \cos \theta_1 \sin \theta_2$$
$$x_j = r \cos \theta_1 \cdots \cos \theta_{j-1} \sin \theta_j, \; j = 1, \ldots, k-2, \; -\frac{\pi}{2} < \theta_j \leq \frac{\pi}{2}$$
$$x_{k-1} = r \cos \theta_1 \cdots \cos \theta_{k-1}, \; -\pi < \theta_{k-1} \leq \pi.$$

Then,

$$dx_1 \wedge \cdots \wedge dx_k = r^{k-1} \left\{ \prod_{j=1}^{k-1} |\cos \theta_j|^{k-1-j} \right\} dr \wedge d\theta_1 \wedge \cdots \wedge d\theta_{k-1}.$$

Now, let us apply Lemma 5.3 to the t_{jk}'s in (5.56). Then, in the real case, $|XX|^\gamma = |TT'|^\gamma = \{\prod_{j=1}^k (t_{jj}^2)^\gamma\}$. The Jacobian part from Lemma 5.2 gives $\prod_{j=1}^k t_{jj}^{q-j}$ and therefore

$$|XX'|^\gamma dX = |TT'|^\gamma dT = \prod_{j=1}^k (t_{jj}^2)^{\gamma+\frac{q}{2}-\frac{j}{2}} dT$$

and hence the r^2 coming from this product is

$$\prod_{j=1}^p (r^2)^{\gamma+\frac{q}{2}-\frac{j}{2}} = (r^2)^{p(\gamma+\frac{q}{2})-p(p+1)/4}. \tag{5.58}$$

The total number of r^2 coming from t_{jk}'s is $p(\gamma + \frac{q}{2}) - \frac{p(p+1)}{4} + \eta + \frac{p(p+1)}{4} - \frac{1}{2} = p(\gamma + \frac{q}{2}) + \eta - \frac{1}{2}$. The product of θ_j's coming from the factor containing $|TT'|$ and the Jacobian part is integrated out in Mathai (2003) and the result, denoted by I_θ, is the following:

$$I_\theta = \frac{2\Gamma_p(\gamma+\frac{q}{2})}{\Gamma(p(\gamma+\frac{q}{2}))}, \ \Re(\gamma) > -\frac{q}{2} + \frac{p-1}{2}. \tag{5.59}$$

Then, the integral in (5.55) is the following for $u = \text{tr} XX'$):

$$M_{a,b}(\delta, \rho; \gamma, \eta) = \int_X |XX'|^\gamma u^\eta e^{-au^\delta - bu^{-\rho}} dX$$

$$= \frac{2\Gamma_p(\gamma+\frac{q}{2})}{\Gamma(p(\gamma+\frac{q}{2}))} \int_{r=0}^\infty (r^2)^{p(\gamma+\frac{q}{2})+\eta-\frac{1}{2}}$$
$$\times e^{-a(r^2)^\delta - b(r^2)^{-\rho}} dr$$

$$= \frac{\Gamma_p(\gamma+\frac{q}{2})}{\Gamma(p(\gamma+\frac{q}{2}))} \int_{y=0}^\infty y^{p(\gamma+\frac{q}{2})+\eta-1}$$
$$\times e^{-ay^\delta - by^{-\rho}} dy. \tag{5.60}$$

Now, the integral in (5.55) is available from the previous section by evaluating the integral over y in (5.60) above.

5.8 Generalized Reaction-Rate Integrals in the Complex Multivariate Case

Now, we look into the extension of reaction-rate integral and general Bessel integral to the complex domain. First we consider the vector case and then the matrix-variate case. Let $\tilde{X}$ be a $p \times 1$ vector in the complex domain with p distinct scalar complex variables as elements $\tilde{x}_j$'s. Then,

$$\tilde{X}^*\tilde{X} = |\tilde{x}_1|^2 + \cdots + |\tilde{x}_p|^2 = (x_{11}^2 + x_{12}^2) + \cdots + (x_{p1}^2 + x_{p2}^2)$$

where $\tilde{x}_j = x_{j1} + ix_{j2}$, $i = \sqrt{(-1)}$, x_{j1}, x_{j2} are real scalar variables. Consider the transformation $y = \tilde{X}^*\tilde{X}$. Note that y here is real also and y contains sum of squares of $2p$ real scalar variables as opposed to p variables in the corresponding real case. Apply Lemma 5.1 to y. Then,

$$d\tilde{X} = \frac{\pi^{\frac{2p}{2}}}{\Gamma(\frac{2p}{2})} y^{\frac{2p}{2}-1} dy = \frac{\pi^p}{\Gamma(p)} y^{p-1} dy. \tag{5.61}$$

Consider the following integral, even though it is real-valued, in order to distinguish it from the corresponding real case, we will use a tilde.

$$\tilde{I}_{a,b}(\delta, \rho) = \int_{\tilde{X}} [\tilde{X}^*\tilde{X}]^\gamma e^{-a(\tilde{X}^*\tilde{X})^\delta - b(\tilde{X}^*\tilde{X})^{-\rho}} d\tilde{X}$$

$$= \frac{\pi^p}{\Gamma(p)} \int_{y=0}^\infty y^{\gamma+p-1} e^{-ay^\delta - by^{-\rho}} dy. \tag{5.62}$$

Now, the integral in (5.62) can be evaluated by using the techniques in Sects. 5.1–5.3. As in the real case, we can consider the pathway extended models also, which can all be evaluated by making the transformation in (5.61) above and then using the techniques from Sects. 5.1–5.3. Hence, we will only list the pathway forms here. For simplicity, we will write them by using the symbol $y = \tilde{X}^*\tilde{X}$ and denoting them by $\tilde{I}_1, \tilde{I}_2, \tilde{I}_3$ for $q_1 > 1, q_2 > 1$:

$$\tilde{I}_1 = \int_{\tilde{X}} y^{\gamma+p-1} [1 + a(q_1-1)y^\delta]^{-\frac{1}{q_1-1}} e^{-by^{-\rho}} d\tilde{X} \tag{5.63}$$

$$\tilde{I}_2 = \int_{\tilde{X}} y^{\gamma+p-1} e^{-ay^\delta} [1 + b(q_2-1)y^{-\rho}]^{-\frac{1}{q_2-1}} d\tilde{X} \tag{5.64}$$

$$\tilde{I}_3 = \int_{\tilde{X}} y^{\gamma+p-1} [1 + a(q_1-1)y^\delta]^{-\frac{1}{q_1-1}} [1 + b(q_2-1)y^{-\rho}]^{-\frac{1}{q_2-1}} d\tilde{X}. \tag{5.65}$$

Observe that there are three situations each in (5.63) and (5.64), that is $q_1 > 1, < 1, \to 1$ and $q_2 > 1, < 1, \to 1$ and there are nine situations in (5.65).

5.9 Complex Matrix-Variate Case

Let $\tilde{X}$ be a $p \times q$, $p \leq q$ matrix of rank p in the complex domain with distinct pq complex scalar variables as elements. Then, $\tilde{Y} = \tilde{X}\tilde{X}^*$ is Hermitian positive definite. Note that

$$\text{tr}(\tilde{X}\tilde{X}^*) = \sum_{j=1}^{p}\sum_{k=1}^{q}|\tilde{x}_{jk}|^2 = \sum_{j=1}^{p}\sum_{k=1}^{q}(x_{jk1}^2 + x_{jk2}^2) \tag{5.66}$$

where $\tilde{x}_{jk} = x_{jk1} + ix_{jk2}$, $i = \sqrt{(-1)}$, x_{jk1}, x_{jk2} are real scalar variables. Thus, there are $2pq$ sum of squares of real scalar variables in (5.66) above, as opposed to pq sum of squares in the corresponding real case. Consider the transformation $y = \text{tr}(\tilde{X}\tilde{X}^*)$. Then, we can apply Lemma 5.1 on a $1 \times 2pq$ real vector and then we have the following result:

$$y = \text{tr}(\tilde{X}\tilde{X}^*) \Rightarrow d\tilde{X} = \frac{\pi^{\frac{2pq}{2}}}{\Gamma(\frac{2pq}{2})} y^{\frac{2pq}{2}-1} dy = \frac{\pi^{pq}}{\Gamma(pq)} y^{pq-1} dy. \tag{5.67}$$

Now, for the $p \times q$, $p \leq q$ matrix of rank p in the complex domain, consider the evaluation of the following integral:

$$\tilde{I}_1 = \int_{\tilde{X}} [\text{tr}(\tilde{X}\tilde{X}^*)]^\gamma e^{-(\text{tr}(\tilde{X}\tilde{X}^*))^\delta - b(\text{tr}(\tilde{X}\tilde{X}^*))^{-\rho}} d\tilde{X}$$

$$= \frac{\pi^{pq}}{\Gamma(pq)} \int_{y=0}^{\infty} y^{\gamma+pq-1} e^{-ay^\delta - by^{-\rho}} dy. \tag{5.68}$$

Now, the integral in (5.68) can be evaluated by using the techniques from Sects. 5.1–5.3.

5.10 Most General Form in the Complex Matrix-Variate Case

Let $\tilde{X} = (\tilde{x}_{jk})$ be a $p \times q$, $p \leq q$ matrix of rank p in the complex domain with pq distinct scalar complex variables as elements. Then $\tilde{X}\tilde{X}^* = \tilde{Y} > O$ (Hermitian positive definite). Consider the integral

$$\tilde{M}_1 = \int_{\tilde{X}} |\det(\tilde{X}\tilde{X}^*)|^\gamma [\text{tr}(\tilde{X}\tilde{X}^*)]^\eta$$

$$\times e^{-a(\text{tr}(\tilde{X}\tilde{X}^*))^\delta - b(\text{tr}(\tilde{X}\tilde{X}^*))^{-\rho}} d\tilde{X}. \tag{5.69}$$

5.10 Most General Form in the Complex Matrix-Variate Case

Consider the transformation $\tilde{X} = \tilde{T}\tilde{U}$, where $\tilde{T}$ is a $p \times p$ lower triangular matrix with real and positive diagonal elements and $\tilde{U}$ is a $p \times q$ unique semi-unitary matrix, $\tilde{U}\tilde{U}^* = I_p$. Then, after integrating out the differential element corresponding to $\tilde{U}$, we have the following result, see also Lemma 5.2:

$$\tilde{X} = \tilde{T}\tilde{U} \Rightarrow d\tilde{X} = \frac{\pi^{pq}}{\tilde{\Gamma}_p(q)} \left\{ \prod_{j=1}^{p} t_{jj}^{2(q-j)+1} \right\} d\tilde{T}. \qquad (5.70)$$

Note that

$$\operatorname{tr}(\tilde{T}\tilde{T}^*) = \sum_{j=1}^{p} t_{jj}^2 + \sum_{j>k} |\tilde{t}_{jk}|^2 \qquad (5.71)$$

where $|\tilde{t}_{jk}|^2 = t_{jk1}^2 + t_{jk2}^2$, $\tilde{t}_{jk} = t_{jk1} + it_{jk2}$, $i = \sqrt{(-1)}$, t_{jk1}, t_{jk2} are real scalar variables. Thus, there is a total of $p + 2\frac{p(p-1)}{2} = p^2$ sum of squares of real scalar variables in $\operatorname{tr}(\tilde{T}\tilde{T}^*)$. In the notation of Lemma 5.3, $k = p^2$ in the complex case and $k = p(p+1)/2$ in the real case. Then, under the general polar coordinate transformation in Lemma 5.3 we have the following, denoting the complex scalar variables as $\tilde{x}_1, \ldots, \tilde{x}_{p^2}$:

$$d\tilde{x}_1 \wedge \cdots \wedge d\tilde{x}_{p^2} = r^{p^2-1} \left\{ \prod_{j=1}^{p^2-1} |\cos\theta_j|^{p^2-j-1} \right\} dr \wedge d\theta_1 \wedge \cdots d\theta_{p^2-1}. \qquad (5.72)$$

and $\operatorname{tr}(\tilde{T}\tilde{T}^*) = r^2$. Sine and cosine of θ_j's are coming from the determinant $|\det(\tilde{T}\tilde{T}^*)|$ and the integral over this function of θ_j's is already evaluated in Mathai (2003) for the real and complex cases. For the complex case, it is the following, denoted by $\tilde{I}_\theta$:

$$\tilde{I}_\theta = \frac{2\tilde{\Gamma}_p(\gamma+q)}{\Gamma(p(\gamma+q))}, \ \Re(\gamma) > -q + p - 1. \qquad (5.73)$$

In the complex case, combining all factors of r^2 we have $(r^2)^{p(\gamma+q)+\eta-\frac{1}{2}}$. Now, the integral in (5.69) reduces to the following:

$$\int_{\tilde{X}} |\det(\tilde{X}\tilde{X}^*)|^\gamma [\operatorname{tr}(\tilde{X}\tilde{X})]^\eta e^{-a(\operatorname{tr}(\tilde{X}\tilde{X}^*))^\delta - b(\operatorname{tr}(\tilde{X}\tilde{X}^*))^{-\rho}} d\tilde{X}$$

$$= \frac{2\tilde{\Gamma}_p(\gamma+q)}{\Gamma(p(\gamma+q))} \int_r (r^2)^{p(\gamma+q)+\eta-\frac{1}{2}} e^{-a(r^2)^\delta - b(r^2)^{-\rho}} dr$$

$$= \frac{\tilde{\Gamma}_p(\gamma+q)}{\Gamma(p(\gamma+q))} \int_{u=0}^{\infty} u^{p(\gamma+q)+\eta-1} e^{-au^\delta - bu^{-\rho}} du. \qquad (5.74)$$

This is the most general form of the integral in the complex case. The evaluation of the integral over u can be done by using the techniques in Sects. 5.1–5.3.

5.11 Reaction-Rate Probability Integral Through Optimization of Mathai Entropy

Basic measure of uncertainty in a scheme (a set of mutually exclusive and totally exhaustive events along with the corresponding probabilities) is Shannon entropy defined as the following, where the first expression is for a discrete distribution and the second item is for a density function $f(x)$ of a real scalar random variable x, both denoted by $S(f)$:

$$S(f) = -c \sum_{j=1}^{k} p_j \ln p_j, \, p_j > 0, \, j = 1, \ldots, k, \, p_1 + \cdots + p_k = 1, \, S(f)$$
$$= -c \int_x f(x) \ln f(x) \mathrm{d}x \tag{5.75}$$

where c is a constant. This basic Shannon entropy is generalized in various directions. One α-generalized entropy is Havrda-Charvat entropy denoted by $H_\alpha(f)$ where

$$H_\alpha(f) = \frac{\int_x [f(x)]^\alpha \mathrm{d}x - 1}{2^{1-\alpha} - 1}, \, \alpha \neq 1. \tag{5.76}$$

A variant of $H_\alpha(f)$ is Tsallis entropy $T_\alpha(f)$ given by

$$T_\alpha(f) = \frac{\int_x [f(x)]^\alpha \mathrm{d}x - 1}{1 - \alpha}, \, \alpha \neq 1. \tag{5.77}$$

Mathai entropy is also a variant of $H_\alpha(f)$, denoted by $M_\alpha(f)$ and defined as

$$M_\alpha(f) = \frac{\int_X [f(X)]^{1 + \frac{a-\alpha}{\eta}} \mathrm{d}X - 1}{\alpha - a}, \, \alpha \neq a, \eta > 0 \tag{5.78}$$

where a is an fixed anchoring point, α is the parameter of interest and the deviation of α from a is measured in η units. Here, $f(X)$ is defined as a real-valued scalar function of X such that $f(X) \geq 0$ for all X in the domain of X and $\int_X f(X) \mathrm{d}X = 1$. It is called a density function where X may be a scalar/vector/matrix or a sequence of matrices in the real or complex domain. Thus, (5.78) is a very general concept. It can also be taken as an expected value, namely

$$M_\alpha(f) = E \left[\frac{[f(X)]^{\frac{a-\alpha}{\eta}} - 1}{\alpha - a} \right] = \frac{E[[f(X)]^{\frac{a-\alpha}{\eta}}] - 1}{\alpha - a}, \, \alpha \neq a, \eta > 0.$$

5.11 Reaction-Rate Probability Integral Through Optimization of Mathai Entropy

Note that when $\alpha \to a$, $M_\alpha(f) \to S(f) =$ Shannon entropy and hence $M_\alpha(f)$ is an α-generalized Shannon entropy. Similarly, $H_\alpha(f) \to S(f)$ and $T_\alpha(f) \to S(f)$ when $\alpha \to 1$. Thus, $H_\alpha(f)$ and $T_\alpha(f)$ are also α-generalized Shannon entropy. Let us consider the optimization of Mathai entropy $M_\alpha(f)$ under the following two moment-type constraints, where we take x as a real scalar variable to start with:

$$(i): E[x^\gamma e^{-b_1 x^{-\rho}}]^{\frac{a-\alpha}{\eta}} = \text{given}; \quad (ii): E[(x^\gamma e^{-b_1 x^{-\rho}})^{\frac{a-\alpha}{\eta}} x^\delta] = \text{given} \quad (5.79)$$

where $b_1 > 0, \rho > 0, \delta > 0, \Re(\gamma) > 0$ and $E[\cdot]$ denotes the expected value of $[\cdot]$. If we use calculus of variation to optimize $M_\alpha(f)$ under the constraints in (5.79) above, then the Euler equation is the following, where λ_1 and λ_2 are the Lagrangian multipliers:

$$\frac{\partial}{\partial f}[f^{1+\frac{a-\alpha}{\eta}} - \lambda_1 (x^\gamma e^{-b_1 x^{-\rho}})^{\frac{a-\alpha}{\eta}} f - \lambda_2 [(x^\gamma e^{-b_1 x^{-\rho}})^{\frac{a-\alpha}{\eta}} x^\delta f]] = 0 \Rightarrow$$

$$f^{\frac{a-\alpha}{\eta}} = \lambda_3 [x^\gamma e^{-b_1 x^{-\rho}}]^{\frac{a-\alpha}{\eta}} [1 + \lambda_4 x^\delta] \Rightarrow$$

$$f = \lambda_5 x^\gamma e^{-b_1 x^{-\rho}} [1 + \lambda_6 x^\delta]^{\frac{\eta}{a-\alpha}}$$

where λ_3 to λ_6 are some constants. Let us consider the case $\alpha < a$. Then, the exponent $\frac{\eta}{a-\alpha} > 0$ since $\eta > 0$. Then, if f has to remain as a density function for all possible values of $a, \alpha, \eta, \alpha < a, \eta > 0$, then λ_6 has to be negative. Let $\lambda_6 = -b(a-\alpha), b > 0, \alpha < a$ and let c_1 be corresponding normalizing constant and let the resulting density be denoted as

$$f_1(x) = c_1 x^\gamma e^{-b_1 x^{-\rho}} [1 - b(a-\alpha) x^\delta]^{\frac{\eta}{a-\alpha}}, \alpha < a, b_1 > 0, b > 0, \eta > 0 \quad (5.80)$$

under the condition $1 - b(a-\alpha) x^\delta > 0, \delta > 0, \rho > 0, \Re(\gamma) > 0$. When $\alpha > a$, then $a - \alpha = -(\alpha - a)$ and then $f_1(x)$ switches into the function

$$f_2(x) = c_2 x^\gamma e^{-b_1 x^{-\rho}} [1 + b(\alpha - a) x^\delta]^{-\frac{\eta}{\alpha-a}}, \alpha > a \quad (5.81)$$

for $b > 0, b_1 > 0, \delta > 0, \rho > 0, \eta > 0, \Re(\gamma) > 0$. When $\alpha \to a$, then both $f_1(x)$ and $f_2(x)$ go to

$$f_3(x) = c_3 x^\gamma e^{-b_1 x^{-\rho}} e^{-b\eta x^\delta} \quad (5.82)$$

for $b_1 > 0, b > 0, \delta > 0, \rho > 0, \eta > 0, \Re(\gamma) > 0$, where c_1, c_2, c_3 are the respective normalizing constants.

5.12 Multivariate Densities in the Real and Complex Domains

Let X be a $p \times 1$ vector of real scalar variables x_j's, $X' = [x_1, \ldots, x_p]$ so that $u = X'X = x_1^2 + \cdots + x_p^2$. If $\tilde{X}$ is a $p \times 1$ vector in the complex domain with $\tilde{X}' = [\tilde{x}_1, \ldots, \tilde{x}_p], \tilde{x}_j = x_{j1} + ix_{j2}, i = \sqrt{(-1)}, x_{j1}, x_{j2}$ are real scalar quantities, $j = 1, \ldots, k$, then, let $\tilde{u} = \tilde{X}^*\tilde{X}$ where $\tilde{X}^*$ is the conjugate transpose of $\tilde{X}$. Then, $\tilde{u} = (x_{11}^2 + x_{12}^2) + \cdots + (x_{p1}^2 + x_{p2}^2) =$ sum of squares of $2p$ real scalar variables when $\tilde{X}$ is in the complex domain. Let X in $f(X)$ of $M_\alpha(f)$ be a $p \times 1$ vector in the real domain. Let us replace x in the constraints (i) and (ii) in (5.79) by u, then $f_j(x)$ of (5.80)–(5.82) change to corresponding densities in terms of X, where for example $f_1(x)$ in the real scalar case changes to

$$f_1(X) = C_1 u^\gamma e^{-b_1 u^{-\rho}} [1 - b(a-\alpha)u^\delta]^{\frac{\eta}{a-\alpha}}, \alpha < a \tag{5.83}$$

for $1 - b(a - \alpha)u^\delta > 0, b > 0, b_1 > 0, \eta > 0, \delta > 0, \rho > 0, \Re(\gamma) > 0, \alpha < a$, $u = X'X$. Then, $f_1(X)$ of (5.83) above is a real multivariate extension of the pathway density for the case $\alpha < a$. We can obtain $f_2(X)$ for $\alpha > a$ and $f_3(X)$ for $\alpha \to a$. Now, let us replace x in the constraints in (i) and (ii) of (5.79) by $\tilde{u} = \tilde{X}^*\tilde{X}$ where $\tilde{X}$ is a $p \times 1$ vector in the complex domain. Then, we will end up with densities corresponding to (5.80)–(5.82), where for example the density for the case $\alpha > a$, denoted by $f_2(\tilde{X})$, is the following:

$$f_2(\tilde{X}) = \tilde{C}_2 \tilde{u}^\gamma e^{-b_1 \tilde{u}^{-\rho}} [1 + b(\alpha - a)\tilde{u}^\delta]^{-\frac{\eta}{\alpha-a}}, \alpha > a \tag{5.84}$$

for $\delta > 0, \rho > 0, b > 0, b_1 > 0, \Re(\gamma) > 0, \eta > 0$.

5.13 Matrix-Variate Densities in the Real and Complex Domains

Let X be a $p \times q, p \leq q$ matrix of rank p in the real domain. Let $u = \text{tr}(XX')$. We use the same notation u for convenience. Observe that $XX' > O$ (positive definite). Let us replace x in the constraints in (i) and (ii) of (5.79) by $u = \text{tr}(XX')$. Then, optimization of Mathai entropy $M_\alpha(f)$ produces three densities denoted by $f_j(X), j = 1, 2, 3$ where X is now a $p \times q, p \leq q$ matrix of rank p in the real domain. For example, $f_2(X)$ in this case will be the following, again denoting the normalizing constant by C_2:

$$f_2(X) = C_2 u^\gamma e^{-b_1 u^{-\rho}} [1 + b(\alpha - a)u^\delta]^{-\frac{\eta}{\alpha-a}}, \alpha > a \tag{5.85}$$

5.14 Mellin Convolutions Involving Other Functions   119

for $b > 0, b_1 > 0, \delta > 0, \rho > 0, \eta > 0, \Re(\gamma) > 0, u = \text{tr}(XX')$. Let the $p \times q$, $p \leq q$ matrix $\tilde{X}$ of rank p be in the complex domain. Let $\tilde{u} = \text{tr}(\tilde{X}\tilde{X}^*)$ where $\tilde{X}^*$ is the conjugate transpose of $\tilde{X}$. Observe that $\tilde{u}$ here is also real and we use the same notation $\tilde{u}$ for convenience. Now, replace x of the constraints in (i) and (ii) of (5.79) by this $\tilde{u}$ and then optimize Mathai entropy $M_\alpha(f)$. Then, we end up with three densities corresponding to (5.80)–(5.82), where for example, $f_3(\tilde{X})$ is the following:

$$f_3(\tilde{X}) = \tilde{C}_3 \tilde{u}^\gamma e^{-b_1 \tilde{u}^{-\rho} - b\eta \tilde{u}^\delta}, \tag{5.86}$$

for $\delta > 0, \rho > 0, b > 0, b_1 > 0, \eta > 0, \Re(\gamma) > 0, \tilde{u} = \text{tr}(\tilde{X}\tilde{X}^*)$. This, $f_3(\tilde{X})$ is the matrix-variate density in the complex domain corresponding to (5.82) in the real scalar case. In a similar manner, one can consider various constraints for the matrix-variate case and obtain all the densities considered in Sects. 5.1–5.4 through optimization of Mathai entropy. In order to limit the size of the manuscript, we stop the discussion of the construction of densities through optimization of entropies.

5.14 Mellin Convolutions Involving Other Functions

In Sects. 5.1–5.7 we have considered reaction-rate probability integral, its generalizations and its extensions. These can also be looked upon as basic Bessel integral, its generalizations and its pathway extensions. They can also be considered as Mellin convolution of a product involving gamma and generalized gamma functions and generalized gamma densities as basic functions. We did not look into Mellin convolution of a ratio involving generalized gamma functions because our interest was to look into generalizations of the basic reaction-rate probability integral. In the present section, we will examine the situation if one of the functions involved is a type-1 beta function type. In terms of statistical densities, we are going to examine Mellin convolution of a product when one of the densities involved is a type-1 beta density. Let $x_1 > 0$, $x_2 > 0$ be two real scalar variables with the associated functions $f_1(x_1)$ and $f_2(x_2)$ respectively and with the joint function $f_1(x_1) f_2(x_2)$, the product, where let

$$f_1(x_1) = c \frac{1}{\Gamma(\alpha)} x_1^\gamma (1-x_1)^{\alpha-1}, 0 \leq x_1 \leq 1 \tag{5.87}$$

for $\Re(\alpha) > 0, \Re(\gamma) > -1$ with $c = \Gamma(\gamma + 1 + \alpha)/\Gamma(\gamma + 1)$ if $f_1(x_1)$ is a statistical density. This density, is the type-1 real scalar beta density with the parameters $(\gamma + 1, \alpha)$ under the usual notation. Let the second function $f_2(x_2) = f(x_2)$ an arbitrary function or arbitrary density if we are considering statistical densities. Let $u = x_1 x_2$. Let the function corresponding to u be $g(u)$. Then, from the Mellin convolution of a product property

$$g(u) = \int_v \frac{1}{v} f_1(\frac{u}{v}) f_2(v) dv = c \frac{1}{\Gamma(\alpha)} (\frac{u}{v})^\gamma (1 - \frac{u}{v})^{\alpha-1} f(v) dv$$

$$= c \frac{u^\gamma}{\Gamma(\alpha)} \int_{v>u>0} v^{-\gamma-\alpha} (v-u)^{\alpha-1} f(v) dv = c \, K_{2,\gamma,u}^{-\alpha}(f) \qquad (5.88)$$

where $K_{2,\gamma,u}^{-\alpha}(f)$ is Erdélyi-Kober fractional integral of the second kind of order α and parameter γ in the real scalar variable case. Thus, a simple change of the basic functions to a type-1 beta form and a general function for $f_2(x_2)$, leads to a fractional integral of the second kind. Mathai (2009) shows that if we take the first function $f_1(x_1)$ as

$$f_1(x_1) = \frac{\phi(x_1)}{\Gamma(\alpha)} (1 - x_1)^{\alpha-1}, \, \Re(\alpha) > 0$$

and $f_2(x_2) = f(x_2)$ as above, then by specializing $\phi(x_1)$, one can obtain all the various definitions of fractional integrals of the second kind of order α introduced by various authors from time to time, in the real scalar case situation. For example, if $\phi(x_1) = x_1^\gamma$ one has Erdélyi-Kober fractional integral. If $\phi(x_1) = x_1^{-\alpha}$, one had Weyl fractional integral of the second kind. In this case, if v is bounded above by a constant b, $v \le b$, then it becomes Riemmann-Liouville fractional integral of the second kind and so on. From the derivation of (5.88), observe that this fractional integral of the second kind can also be obtained as a constant multiple of a density of a product of real scalar positive random variables when one of the densities is a type-1 beta density with the parameters $(\gamma + 1, \alpha)$ and the other is an arbitrary density.

5.15 Generalization to Real Matrix-Variate Case

Let $X_1 > O$ and $X_2 > O$ be two $p \times p$ real positive definite matrices with distinct elements as real scalar variables. Let the associated functions be the real-valued scalar functions $f_1(X_1)$ and $f_2(X_2)$ respectively and let the joint function be $f_1(X_1) f_2(X_2)$. If X_1 and X_2 are matrix-variate real random variables with the associated densities $f_1(X_1)$ and $f_2(X_2)$ then, when the joint function is the product $f_1(X_1) f_2(X_2)$, we say that X_1 and X_2 are statistically independently distributed. Consider the symmetric product $U = X_2^{\frac{1}{2}} X_1 X_2^{\frac{1}{2}}$ and let $X_2 = V$ where $X_2^{\frac{1}{2}}$ is the symmetric positive definite square root of the positive definite matrix $X_2 > O$. From Mathai (1997) we have the Jacobian and we can show that $dX_1 \wedge dX_2 = |V|^{-\frac{p+1}{2}} dU \wedge dV$. This involves the transformation of a symmetric matrix to a symmetric matrix and we have not listed the corresponding Jacobian as a lemma in order to save space. Then, the function corresponding to U, again denoted by $g(U)$, is the following:

$$g(U) = \int_V |V|^{-\frac{p+1}{2}} f_1(V^{-\frac{1}{2}} U V^{-\frac{1}{2}})^\gamma f_2(V) dV. \qquad (5.89)$$

Let

$$f_1(X_1) = \frac{1}{\Gamma_p(\alpha)}|X_1|^\gamma|I - X_1|^{\alpha - \frac{p+1}{2}}, O < X_1 < I, \Re(\alpha) > \frac{p-1}{2} \quad (5.90)$$

and let $f_2(X_2) = f(X_2)$ where f is an arbitrary function, $O < X_1 < I$ means $X_1 > O, I - X_1 > O$ (both positive definite), and $f_1(X_1) = 0$ elsewhere. As observed in the real scalar case, if we multiply by the constant $\Gamma_p(\gamma + \frac{p+1}{2} + \alpha)/\Gamma_p(\gamma + \frac{p+1}{2})$, then $f_1(X_1)$ becomes a real matrix-variate type-1 beta density. Now, from (5.89) and (5.90),

$$g(U) = \frac{1}{\Gamma_p(\alpha)} \int_V |V|^{-\frac{p+1}{2}} |(V^{-\frac{1}{2}}UV^{-\frac{1}{2}})|^\gamma |I - V^{-\frac{1}{2}}UV^{-\frac{1}{2}}|^{\alpha - \frac{p+1}{2}} f(V) dV$$

$$= \frac{|U|^\gamma}{\Gamma_p(\alpha)} \int_{V-U>O} |V|^{-\alpha-\gamma}|V - U|^{\alpha - \frac{p+1}{2}} f(V) dV$$

$$= K_{2,\gamma,U}^{-\alpha}(f) \quad (5.91)$$

where $K_{2,\gamma,U}^{-\alpha}(f)$ is Erdélyi-Kober fractional integral of the second kind of order α and parameter γ in the real matrix-variate case. Since (5.91) is identical with Erdélyi-Kober fractional integral of the second kind when $p = 1$ or in the real scalar variable case, Mathai (2015) defined (5.91) as Erdélyi-Kober fractional integral of the second kind in the real matrix-variate case. Again, in the real matrix-variate case also, if we take $f_1(X_1) = \frac{1}{\Gamma_p(\alpha)}\phi(X_1)|I - X_1|^{\alpha - \frac{p+1}{2}}$ for $O < X_1 < I$, then we can show that by specializing $\phi(X_1)$ one can obtain the real matrix-variate versions of the various definitions of fractional integral of the second kind introduced by various authors for the real scalar variable case. These, matrix-variate versions are not defined by anyone so far.

5.16 Generalization to the Complex Matrix-Variate Case

Let $\tilde{X}_1 > O$ and $\tilde{X}_2 > O$ be two $p \times p$ Hermitian positive definite matrices in the complex domain with the associated functions $f_1(\tilde{X}_1)$ and $f_2(\tilde{X}_2)$ respectively and with the joint unction $f_1(\tilde{X}_1)f_2(\tilde{X}_2)$ where f_1 and f_2 are real-valued scalar functions of the matrix argument in the complex domain. Let $\tilde{U} = \tilde{X}_2^{\frac{1}{2}}\tilde{X}_1\tilde{X}_2^{\frac{1}{2}}$ and $\tilde{V} = \tilde{X}_2$, where $\tilde{X}_2^{\frac{1}{2}}$ is the Hermitian positive definite square root of the Hermitian positive definite matrix $\tilde{X}_2$. Again, from Mathai (1997) we can show that $d\tilde{X}_1 \wedge d\tilde{X}_2 = |\det(\tilde{V})|^{-p} d\tilde{U} \wedge d\tilde{V}$. Let

$$f_1(\tilde{X}_1) = \frac{1}{\tilde{\Gamma}_p(\alpha)}|\det(\tilde{X}_1)|^\gamma|\det(I - \tilde{X}_1)|^{\alpha - p}, O < \tilde{X}_1 < I, \Re(\alpha) > p - 1$$

and $f_1(\tilde{X}_1) = 0$ elsewhere, and let $f_2(\tilde{X}_2) = f(\tilde{X}_2)$ an arbitrary function. Then,

$$g(\tilde{U}) = \frac{1}{\tilde{\Gamma}_p(\alpha)} \int_{\tilde{V}} |\det(\tilde{V})|^{-p} |\det(\tilde{V}^{-\frac{1}{2}} \tilde{U} \tilde{V}^{-\frac{1}{2}})|^{\gamma}$$
$$\times |\det(I - \tilde{V}^{-\frac{1}{2}} \tilde{U} \tilde{V}^{-\frac{1}{2}})|^{\alpha - p} f(\tilde{V}) d\tilde{V}. \qquad (5.92)$$
$$= \frac{|\det(\tilde{U})|^{\gamma}}{\tilde{\Gamma}_p(\alpha)} \int_{\tilde{V} - \tilde{U} > O} |\det(\tilde{V})|^{-\gamma - \alpha} |\det(\tilde{V} - \tilde{U})|^{\alpha - p} f(\tilde{V}) d\tilde{V} \qquad (5.93)$$
$$= \tilde{K}_{2,\gamma,\tilde{U}}^{-\alpha}(f) \qquad (5.94)$$

where $\tilde{K}_{2,\gamma,\tilde{U}}^{-\alpha}(f)$ is Erdélyi-Kober fractional integral of the second kind of order α and parameter γ in the complex matrix-variate case. Mathai (2013) has defined like this when he extended fractional calculus to the complex matrix-variate case. In the real scalar case (5.94) will agree with Erdélyi-Kober fractional integral of the second kind.

5.17 Mellin Convolution of a Ratio

Let us start with real scalar variables first. Let $x_1 > 0$ and $x_2 > 0$ be two real scalar positive variables. We can define a product in two different ways $u = x_1 x_2$, $v = x_2$ or $v = x_1$. We can define a ratio in four different ways $u_1 = \frac{x_1}{x_2}$ with $v = x_1$ or $v = x_2$ and $u_1 = \frac{x_2}{x_1}$ with $v = x_1$ or $v = x_2$. All these different representations can lead to different items in different disciplines as can be seen from the following illustration. Let $u_1 = \frac{x_2}{x_1}$ with $v = x_2$ so that

$$x_2 = v, \ x_1 = \frac{v}{u_1} \Rightarrow dx_1 \wedge dx_2 = -\frac{v}{u_1^2} du_1 \wedge dv.$$

Let the function corresponding to u_1 be denoted by $g_1(u_1)$. Then,

$$g_1(u_1) = \int_v \frac{v}{u_1^2} f_1(\frac{v}{u_1}) f_2(v) dv. \qquad (5.95)$$

Let

$$f_1(x_1) = \frac{x_1^{\gamma - 1}}{\Gamma(\alpha)} (1 - x_1)^{\alpha - 1}, 0 < x_1 < 1, \Re(\alpha) > 0, \Re(\gamma) > 0$$

and $f_2(x_2) = f(x_2)$ an arbitrary function. Then,

5.18 M-Convolution of a Ratio in the Real Matrix-Variate Case

$$g_1(u_1) = \frac{1}{\Gamma(\alpha)} \int_v \frac{v}{u_1^2} (\frac{v}{u_1})^{\gamma-1} (1 - \frac{v}{u_1})^{\alpha-1} f(v) v$$

$$= \frac{u_1^{-\alpha-\gamma}}{\Gamma(\alpha)} \int_{0<v<u_1} v^\gamma (u_1 - v)^{\alpha-1} f(v) dv$$

$$= K_{1,\gamma,u_1}^{-\alpha}(f) \tag{5.96}$$

where $K_{1,\gamma,u_1}^{-\alpha}(f)$ is Erdélyi-Kober fractional integral of the first kind of order α and parameter γ for the real scalar variable case. By multiplying f_1 with a constant, one has a real scalar type-1 beta density and if $f(x_2)$ is an arbitrary density, then the fractional integral of the first kind in (ii) is a constant multiple of the density of a ratio of two independently distributed real scalar positive random variables. This connection was pointed out by Mathai (2009).

5.18 M-Convolution of a Ratio in the Real Matrix-Variate Case

Let $X_1 > O$ and $X_2 > O$ be two $p \times p$ real positive definite matrices and let the associated functions be $f_1(X_1)$ and $f_2(X_2)$ respectively and let the joint function be $f_1(X_1) f_2(X_2)$. Let $U_1 = X_2^{\frac{1}{2}} X_1^{-1} X_2^{\frac{1}{2}}$ be a symmetric ratio of the matrices X_1 and X_2. Let $V = X_2$. Then, from Mathai (1997), we can show that

$$dX_1 \wedge dX_2 = |U_1|^{-(p+1)} |V|^{\frac{p+1}{2}} dU_1 \wedge dV.$$

Let

$$f_1(X_1) = \frac{1}{\Gamma_p(\alpha)} |X_1|^{\gamma - \frac{p+1}{2}} |I - X_1|^{\alpha - \frac{p+1}{2}}, O < X_1 < I,$$

for $\Re(\alpha) > \frac{p-1}{2}, \Re(\gamma) > \frac{p-1}{2}$ and let $f_2(X_2) = f(X_2)$ an arbitrary function. Then,

$$g_1(U_1) = \int_V |U_1|^{-(p+1)} |V|^{\frac{p+1}{2}} f_1(V^{\frac{1}{2}} U_1^{-1} V^{\frac{1}{2}}) f(V) dV$$

$$= \frac{1}{\Gamma_p(\alpha)} \int_V |U_1|^{-(p+1)} |V|^{\frac{p+1}{2}} |V^{\frac{1}{2}} U_1^{-1} V^{\frac{1}{2}}|^{\gamma - \frac{p+1}{2}}$$

$$\times |I - V^{\frac{1}{2}} U_1^{-1} V^{\frac{1}{2}}|^{\alpha - \frac{p+1}{2}} f(V) dV$$

$$= \frac{|U_1|^{-\gamma-\alpha}}{\Gamma_p(\alpha)} \int_{U_1 - V > O} |V|^\gamma |U_1 - V|^{\alpha - \frac{p+1}{2}} f(V) dV$$

$$= K_{1,\gamma,U_1}^{-\alpha}(f) \tag{5.97}$$

where $K_{1,\gamma,U_1}^{-\alpha}(f)$ is Erdélyi-Kober fractional integral of the first kind of order α and parameter γ in the real matrix-variate case. This was established by Mathai in one of the series of papers published in Linear Algebra and Its Applications in 2013–2016, extending fractional calculus to the matrix-variate cases in the real and complex domain.

5.19 Fractional Integral of the First Kind in the Complex Matrix-Variate Case

Let $\tilde{X}_1 > O$ and $\tilde{X}_2 > O$ be $p \times p$ Hermitian positive definite matrices in the complex domain with the associated functions $f_1(\tilde{X}_1)$ and $f_2(\tilde{X}_2)$ respectively and with the joint function $f_1(\tilde{X}_1) f_2(\tilde{X}_2)$ where f_1 and f_2 are real-valued scalar functions of the matrices $\tilde{X}_1$ and $\tilde{X}_2$ in the complex domain. Consider the symmetric ratio $\tilde{U}_1 = \tilde{X}_2^{\frac{1}{2}} \tilde{X}_1^{-1} \tilde{X}_2^{\frac{1}{2}}$ and $\tilde{V} = \tilde{X}_2$ where $\tilde{X}_2^{\frac{1}{2}}$ is the Hermitian positive definite square root of the Hermitian positive definite matrix $\tilde{X}_2 > O$. From Mathai (1997), it can be shown that $d\tilde{X}_1 \wedge d\tilde{X}_2 = |\det(\tilde{U}_1)|^{-2p} |\det(\tilde{V})|^p d\tilde{U}_1 \wedge d\tilde{V}$ Let

$$f_1(\tilde{X}_1) = \frac{1}{\tilde{\Gamma}_p(\alpha)} |\det(\tilde{X}_1)|^{\gamma-p} |\det(I - \tilde{X}_1)|^{\alpha-p}, O < \tilde{X}_1 < I$$

for $\Re(\alpha) > p-1, \Re(\gamma) > p-1$ and $O < \tilde{X}_1 < I$ means $\tilde{X}_1 > O, I - \tilde{X}_1 > O$ (both Hermitian positive definite). Let $f_2(\tilde{X}_2) = f(\tilde{X}_2)$, an arbitrary function. Then for $\Re(\alpha) > p-1$, the function corresponding to $\tilde{U}_1$, denoted by $g_1(\tilde{U}_1)$, is the following:

$$\begin{aligned}
g_1(\tilde{U}_1) &= \frac{1}{\tilde{\Gamma}_p(\alpha)} \int_{\tilde{V}} |\det(\tilde{V})|^p |\det(\tilde{U}_1)|^{-2p} |\det(\tilde{V}^{\frac{1}{2}} \tilde{U}_1^{-1} \tilde{V}^{\frac{1}{2}})|^{\gamma-p} \\
&\quad \times |\det(I - \tilde{V}^{\frac{1}{2}} \tilde{U}_1^{-1} \tilde{V}^{\frac{1}{2}})|^{\alpha-p} f(\tilde{V}) d\tilde{V} \\
&= \frac{\det(\tilde{U}_1)|^{-\gamma-\alpha}}{\tilde{\Gamma}_p(\alpha)} \int_{(\tilde{U}_1 - \tilde{V}) > O} |\det(\tilde{V})|^{\gamma} |\det(\tilde{U}_1 - \tilde{V})|^{\alpha-p} f(\tilde{V}) dV \\
&= \tilde{K}_{1,\gamma,\tilde{U}_1}^{-\alpha}(f) \qquad (5.98)
\end{aligned}$$

where $\tilde{K}_{1,\gamma,\tilde{U}_1}^{-\alpha}(f)$ is Erdélyi-Kober fractional integral of the first kind of order α and parameter γ in the complex domain as defined by Mathai (2013). A brief overview of the development of fractional calculus in the complex domain and for functions of matrix argument is given in the Springer Brief of Mathai and Haubold (2018). Hence, further discussion of fractional integrals and fractional derivatives is not attempted here.

References

Mathai, A.M.: A Handbook of Generalized Special Functions for Statistical and Physical Sciences. Oxford University Press, Oxford (1993)

Mathai, A.M.: Jacobians of Matrix Transformations and Functions of Matrix Argument. World Scientific Publishing, New York (1997)

Mathai, A.M.: An Introduction to Geometrical Probability: Distributional Aspects with Applications. Gordon and Breach, Amsterdam (1999)

Mathai, A.M.: A pathway to matrix-variate gamma and normal densities. Linear Algebr. Appl. **396**, 317–328 (2005)

Mathai, A.M.: Fractional integrals in the matrix-variate case and connection to statistical distributions. Integr. Transform. Spec. Funct. **20**(12), 871–882 (2009)

Mathai, A.M.: Generalized Krätzel integral and associated statistical distributions. Int. J. Math. Anal. **65**(51), 2501–2510 (2012)

Mathai, A.M.: Fractional integral operators in the complex matrix-variate case. Linear Algebr. Appl. **439**, 2901–2913 (2013). https://doi.org/10.1016/j.laa.2013.08.023

Mathai, A.M.: Fractional integral operators involving many matrix variables. Linear Algebr. Appl. **446**, 196–215 (2014)

Mathai, A.M.: Fractional differential operators in the complex matrix-variate case. Linear Algebr. Appl. **478**, 200–217 (2015)

Mathai, A.M.: Computational representation of the ultra gamma integral. J. Ramanujan Soc. Math. Math. Sci. **5**(2), 39–46 (2016)

Mathai, A.M.: Mellin convolutions, statistical distributions and fractional calculus. Fract. Calc. Appl. Anal. **21**(2), 376–398 (2018). https://doi.org/10.1515/fca-2018-0022

Mathai, A.M.: Some matrix-variate models applicable in different areas. Axioms **12**, 931 (2023). https://doi.org/10.3390/axioms12100.936

Mathai, A.M., Haubold, H.J.: Modern Problems in Nuclear and Neutrino Astrophysics. Akademie-Verlag, Berlin (1988)

Mathai, A.M., Haubold, H.J.: Erdélyi-Kober Fractional Integrals from a Statistical Perspective. Inspired by Neutrino Astrophysics, Springer Nature, Japan (2018)

Mathai, A.M., Haubold, H.J.: A versatile integral in astronomy and astrophysics. Axioms **618275**, 8,122 (2019). https://doi.org/10.3390/axioms.8040/22

Mathai, A.M., Provost, S.B., Haubold, H.J.: Multivariate Statistical Analysis in the Real and Complex Domains. Springer Nature, Switzerland (2022)

Mathai, A.M., Saxena, R.K., Haubold, H.J.: The H-function: Theory and Applications. Springer, New York (2010)

Open Access This chapter is licensed under the terms of the Creative Commons Attribution 4.0 International License (http://creativecommons.org/licenses/by/4.0/), which permits use, sharing, adaptation, distribution and reproduction in any medium or format, as long as you give appropriate credit to the original author(s) and the source, provide a link to the Creative Commons license and indicate if changes were made.

The images or other third party material in this chapter are included in the chapter's Creative Commons license, unless indicated otherwise in a credit line to the material. If material is not included in the chapter's Creative Commons license and your intended use is not permitted by statutory regulation or exceeds the permitted use, you will need to obtain permission directly from the copyright holder.

Chapter 6
Neutrino Astrophysics, 2025 Update: The Entropic Approach to Solar Neutrinos

6.1 Solar Neutrinos: SuperKamiokande Data

Over the past 50 years, radio-chemical and real-time solar neutrino experiments have proven to be sensitive tools to test both astrophysical and elementary particle physics models and principles (Sakurai 2018; Orebi Gann et al. 2021). Solar neutrino detectors (radio-chemical: Homestake, GALLEX + GNO, SAGE, real-time: SuperKamiokande, SNO, Borexino) have demonstrated that the Sun is powered by thermonuclear fusion reactions. Today fluxes, particularly from the pp-chain have been measured: pp, ^7Be, pep, ^8B, and, hep. Experiments with solar neutrinos and reactor anti-neutrinos (KamLAND) have confirmed that solar neutrinos undergo flavor oscillations (Mikheyev–Smirnov–Wolfenstein (MSW) model). Results from solar neutrino experiments are consistent with the Mikheyev–Smirnov–Wolfenstein Large Mixing Angle (MSW-LMA) model, which predicts a transition from vacuum-dominated to matter-enhanced oscillations, resulting in an energy dependent electron neutrino survival probability.

6.2 Diffusion Entropy and Standard Deviation: Analysis

For all radio-chemical and real-time solar neutrino experiments, periodic variation in the detected solar neutrino fluxes have been reported, based mainly on Fourier and wavelet analysis methods (standard deviation analysis). Other attempts to analyze the same data sets, particularly undertaken by the experimental collaborations of real-time solar neutrino experiments themselves, have failed to find evidence for such variations of the solar neutrino flux over time (Abe et al. 2024a). Periodicities in the solar neutrino fluxes, if confirmed, could provide evidence for new solar, nuclear, or neutrino physics beyond the commonly accepted physics of vacuum-dominated and matter-enhanced oscillations of massive neutrinos (MSW model) that is, after 50

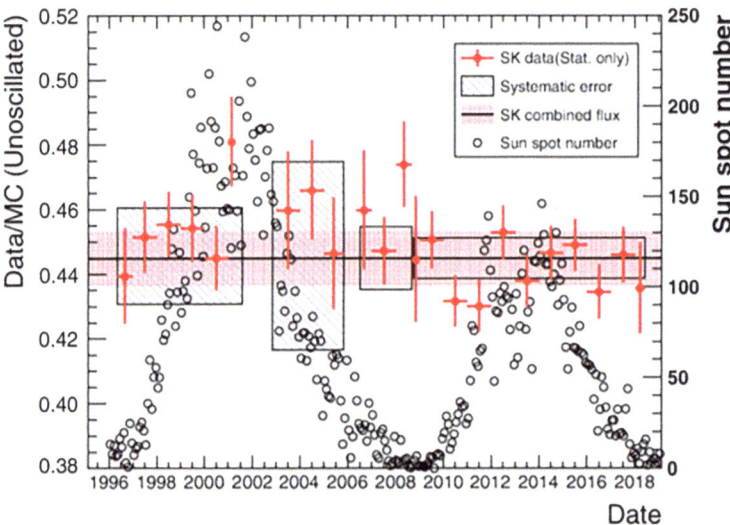

Fig. 6.1 Yearly solar neutrino flux measured by SuperKamiokande. The redfilled circle points show the SuperKamiokande data with statistical uncertainty and the gray striped area show the systematic uncertainty for each phase. The horizontal black solid line (red shaded area) shows the combined value of measured flux (its combined uncertainty). The black-blank circle points show the sunspot numbers from 1996 to 2018 (Abe et al. 2024b)

years of solar neutrino experiment and theory, considered to be the ultimate solution to the solar neutrino problem (Fig. 6.1).

Specifically, subsequent to the analysis made by the SuperKamiokande collaboration, the SNO experiment collaboration has painstakingly searched for evidence of time variability at periods ranging from 10 years down to 10 min. SNO has found no indications for any time variability of the ^8B flux at any timescale, including in the frequency window in which g-mode oscillations of the solar core might be expected to occur. Despite large efforts to utilize helio-seismology and helio-neutrinospectroscopy, at present time there is no conclusive evidence in terms of physics for time variability of the solar neutrino fluxes from any solar neutrino experiment. If such a variability over time would be discovered, a mechanism for a chronometer for solar variability could be proposed based on relations between properties of thermonuclear fusion and g-modes (Buldgen et al. 2024; Sturrock et al. 2021).

All above findings encouraged the conclusion that Fourier and wavelet analysis, which are based upon the analysis of the variance of the respective time series (standard deviation analysis: SDA) should be complemented by the utilization of diffusion entropy analysis (DEA), which measures the scaling of the probability density function (pdf) of the diffusion process generated by the time series thought of as the physical source of fluctuations (Scafetta 2010). For this analysis, we have used the publicly available data of SuperKamiokande-I (1996-05-31–2001-07-15) and SuperKamiokande-II (2002-12-10–2005-10-06) (see Fig. 6.2) (Yoo et al. 2003:

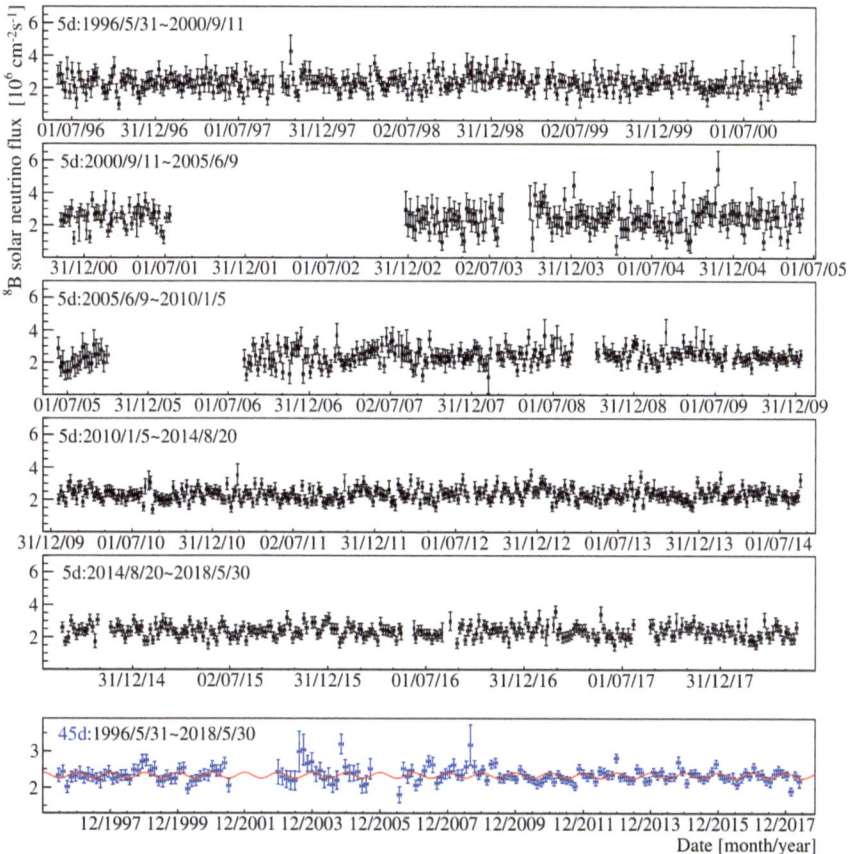

Fig. 6.2 Measured ^8B solar neutrino fuxes for 5-day (top five panels, black data points) and 45-day (bottom panel, blue data points) intervals without $1/R^2$ correction. The errors in the 5-day (the 45-day) plot are asymmetric (symmetric) errors of the average fluxes. The solid-red curve in the 45-day plot is the expected sinusoidal solar neutrino flux based on the elliptical orbit of the Earth (Abe et al. 2024a)

Cravens et al. 2008; Abe et al. 2024b). Such an analysis does not reveal periodic variations of the solar neutrino fluxes but shows how the pdf scaling exponent departs in the non-Gaussian case from the Hurst exponent. Figures 6.3 and 6.4 show the scaling exponents (DEA) for the SuperKamiokande I and II data. The respective Hurst exponents for SDA are visible in Figs. 6.5 and 6.6 (Mathai and Haubold 2018). SuperKamiokande is sensitive mostly to neutrinos from the ^8B and *hep* branch of the *pp* nuclear fusion chain in solar burning. Above approximately 4 MeV the detector can pick-out the scattering of solar neutrinos off atomic electrons which produces Cherenkov radiation in the detector. The ^8B and rarer *hep* neutrinos have a spectrum which ends near 20 MeV.

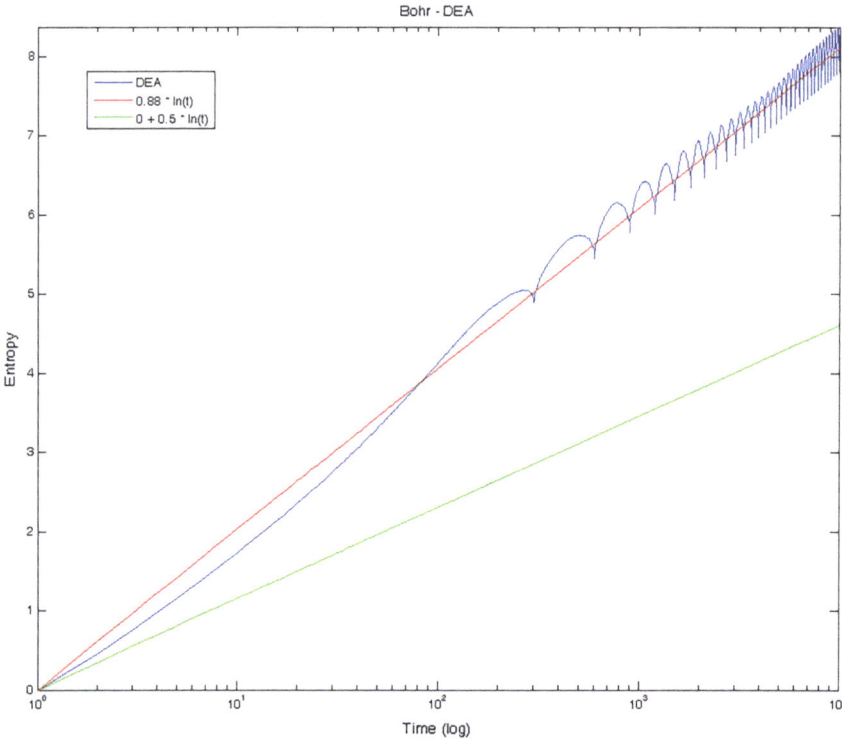

Fig. 6.3 The Diffusion Entropy Analysis (DEA) of the ^8B solar neutrino data from the SuperKamiokande I and II experiment

Assuming that the solar neutrino signal is governed by a probability density function with scaling given by the asymptotic time evolution of a pdf of x, obeying the property (Scafetta 2010; Culbreth et al. 2023)

$$p(x,t) = \frac{1}{t^\delta} f\left(\frac{x}{t^\delta}\right), \tag{6.1}$$

where δ denotes the scaling exponent of the pdf. In the variance based methods, scaling is studied by direct evaluation of the time behavior of the variance of the diffusion process. If the variance scales, one would have

$$\sigma_x^2(t) \sim t^{2H}, \tag{6.2}$$

where $\sigma_x^2(t)$ is the variance of the diffusion process and where H is the Hurst exponent. To evaluate the Shannon entropy of the diffusion process at time t, defined $S(t)$ as

$$S(t) = -\int_{-\infty}^{+\infty} dx \, p(x,t) \ln p(x,t) \tag{6.3}$$

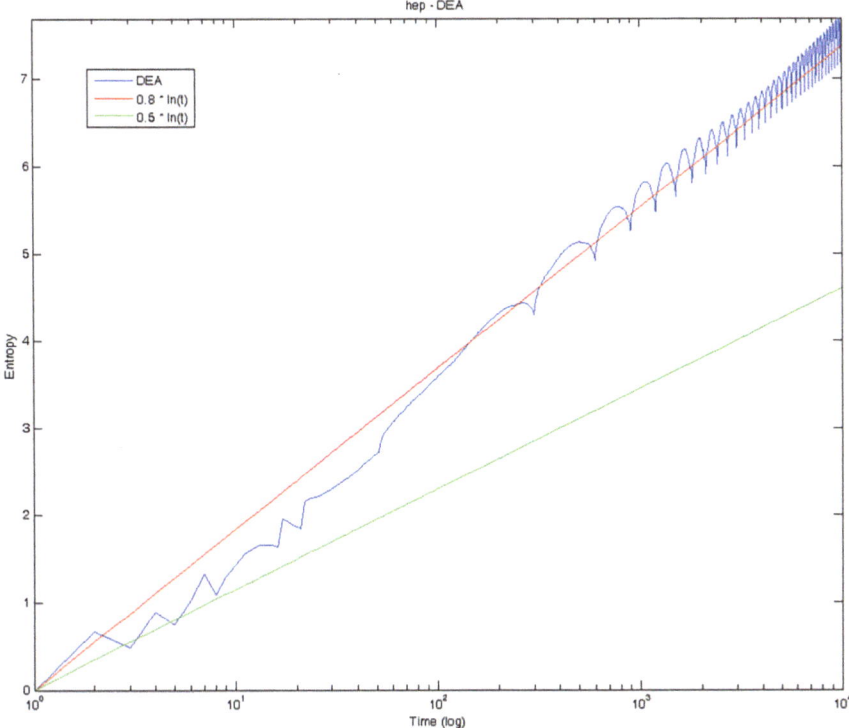

Fig. 6.4 The Diffusion Entropy Analysis (DEA) of the *hep* solar neutrino data from the SuperKamiokande I and II experiment

and with the previous $p(x, t)$ one has

$$S(t) = A + \delta \ln(t), \quad A = -\int_{-\infty}^{+\infty} dy f(y) \ln f(y). \tag{6.4}$$

The scaling exponent δ is the slope of the entropy against the logarithmic time scale. The slope is visible in Figs. 6.3 and 6.4 for the SuperKamiokande data measured for ^8B and *hep*. The Hurst exponents (SDA) are $H = 0.66$ and $H = 0.36$ for ^8B and *hep*, respectively, see Figs. 6.5 and 6.6 (Mathai and Haubold 2018). The *pdf* scaling exponents (DEA) are $\delta = 0.88$ and $\delta = 0.80$ for ^8B and *hep*, respectively, as shown in Figs. 6.3 and 6.4. The values for both SDA and DEA indicate a deviation from Gaussian behavior which would require that $H = \delta = 0.5$.

A test computation for the application of SDA and DEA to data that are known to exhibit non-Gaussian behavior have been published by Haubold et al. (2012) and Tsallis (2024). In this test computation, SDA and DEA, applied to the magnetic field strength fluctuations recorded by the Voyager-I spacecraft in the heliosphere clearly revealed the scaling behavior of such fluctuations as previously already discovered

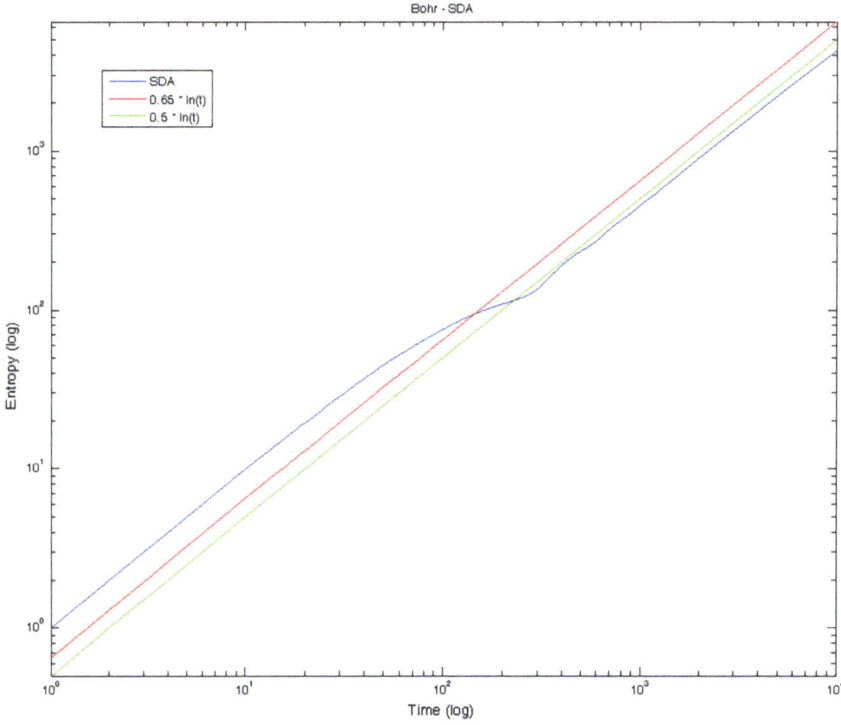

Fig. 6.5 The Standard Deviation Analysis (SDA) of the ^8B solar neutrino data from the SuperKamiokande I and II experiment

by non-extensive statistical mechanics considerations that lead to the determination of the non-extensivity q-triplet.

6.3 Probability Density Function and Differential Equation: Lévy Flights

We consider a diffusion process generated by a waiting time pdf with a finite characteristic time T that can be modeled with a Poissonian distribution, and a jump length pdf $\lambda(x)$ given by a Lévy distribution with index $0 < \alpha < 2$ (Metzler and Klafter 2000). The Fourier transform of $\lambda(x)$ is

$$\hat{\lambda}(k) = exp(-\sigma^\alpha |k|^\alpha) \sim 1 - \sigma^\alpha |k|^\alpha. \tag{6.5}$$

Then $\lambda(x)$ has the asymptotic behavior given by

$$\lambda(x) \sim A_\alpha \sigma^\alpha |x|^{-1-\alpha} = A_\alpha \sigma^{1-\mu} |x|^{-\mu} \tag{6.6}$$

6.3 Probability Density Function and Differential Equation: Lévy Flights 133

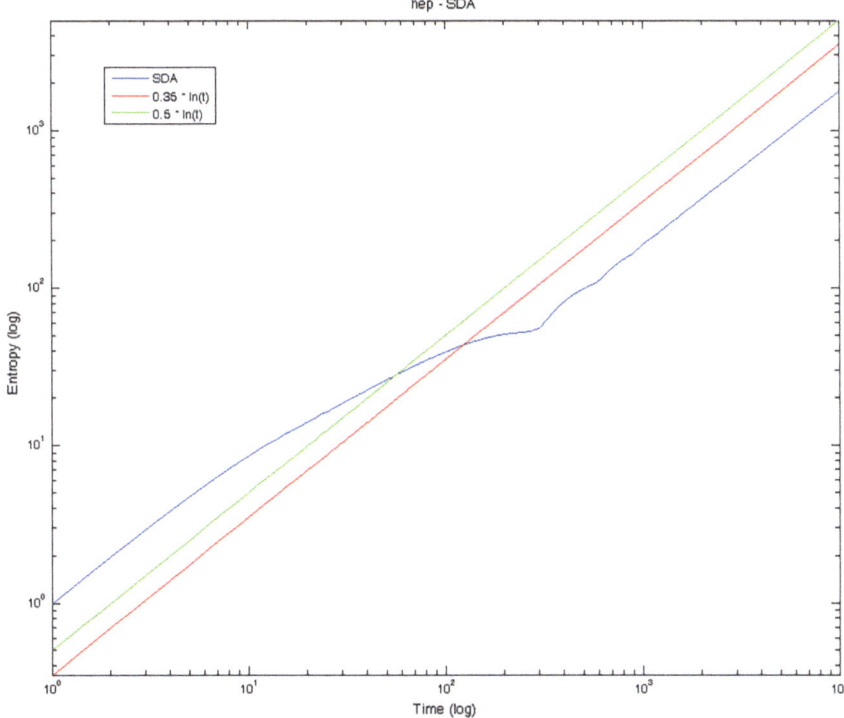

Fig. 6.6 The Standard Deviation Analysis (SDA) of the *hep* solar neutrino data from the SuperKamiokande I and II experiment

for $|x| \gg \sigma$ and $\mu = 1 + \alpha$. Substituting the asymptotic expansion of the jump length pdf $\hat{\lambda}(k)$ in the Fourier space and the waiting time pdf of

$$\Phi(t) = \frac{1}{\tau} exp\left(-\frac{t}{\tau}\right) \tag{6.7}$$

where $\tau = T < \infty$ is the characteristic waiting time in the Laplace space into

$$\hat{p}(k,s) = \frac{1 - \hat{\Phi}(s)}{s} \frac{\hat{p}_0(k)}{1 - \hat{\Phi}(s)\hat{\lambda}(k)}, \tag{6.8}$$

where $\hat{p}_0(k)$ is the Fourier transform of the initial condition $p(x, 0)$, we obtain the following jump pdf in the Fourier-Laplace space

$$\hat{p}(k,s) = \frac{1}{s + K^\alpha |k|^\alpha}, \tag{6.9}$$

where $K^\alpha = \sigma^\alpha/\tau$ is the generalized diffusion constant. Equation (6.9) is the solution of the generalized diffusion equation (Hilfer 2018)

$$\frac{\partial p(x,t)}{\partial t} = K^\alpha {}_{-\infty}D_x^\alpha p(x,t), \qquad (6.10)$$

where ${}_{-\infty}D_x^\alpha$ is the fractional Weyl operator. Upon Laplace inversion of Eq. (6.9), we get the characteristic function of the jump pdf

$$\hat{p}(k,t) = exp(-K^\alpha t |k|^\alpha). \qquad (6.11)$$

Equation (6.11) is the characteristic function of a centered and symmetric Lévy distribution. The Fourier inversion of (6.7) can be analytically obtained by making use of the Fox function (Mathai et al. 2010; Mathai and Haubold 2018)

$$\begin{aligned}p(x,t) &= \frac{1}{\alpha}\frac{1}{t^{1/\alpha}}\frac{t^{1/\alpha}}{|x|} H_{2,2}^{1,1}\left[\frac{|x|}{Kt^{1/\alpha}}\Big|_{(1,1),(1,\frac{1}{2})}^{(1,\frac{1}{\alpha}),(1,\frac{1}{2})}\right] \\ &= \frac{1}{\alpha|x|}\frac{1}{2\pi i}\int_{c-i\infty}^{c+i\infty} \frac{\Gamma(1+\frac{s}{\alpha})\Gamma(-\frac{s}{2})}{\Gamma(-s)\Gamma(1+\frac{s}{2})}\left(\frac{|x|}{Kt^{1/\alpha}}\right)^{-s} ds,\end{aligned}$$

where c in the contour is such that $-\alpha < c < 0$. Replacing $\frac{s}{\alpha}$ by s, observing that α coming from $s\,ds$ is canceled with α sitting outside, we have the following:

$$p(x,t) = \frac{1}{|x|}\frac{1}{2\pi i}\int_{c'-i\infty}^{c'+i\infty} \frac{\Gamma(1+s)\Gamma(-\frac{\alpha}{2}s)}{\Gamma(-\alpha s)\Gamma(1+\frac{\alpha}{2}s)}\left(\frac{|x|}{Kt^{1/\alpha}}\right)^{-\alpha s} ds,\; -1 < c' < 0.$$

By using the duplication formula for gamma functions, we have

$$\Gamma(-\alpha s) = \Gamma\left(2\left(-\frac{\alpha}{2}s\right)\right) = 2^{-\alpha s - 1}\pi^{-\frac{1}{2}}\Gamma\left(-\frac{\alpha}{2}s\right)\Gamma\left(\frac{1}{2} - \frac{\alpha}{2}s\right)$$

so that one $\Gamma(-\frac{\alpha}{2}s)$ is canceled. Then,

$$p(x,t) = \frac{2\pi^{\frac{1}{2}}}{|x|}\frac{1}{2\pi i}\int_{c'-i\infty}^{c'+i\infty} \frac{\Gamma(1+s)}{\Gamma\left(\frac{1}{2} - \frac{\alpha}{2}s\right)\Gamma\left(1 + \frac{\alpha}{2}s\right)}\left(\frac{|x|}{2Kt^{1/\alpha}}\right)^{-\alpha s} ds.$$

Evaluating the H-function as the sum of the residues at the poles of $\Gamma(1+s)$, which are at $s = -1 - \nu$, $\nu = 0, 1, \ldots$, we have the following series:

$$\begin{aligned}p(x,t) = &\frac{2\pi^{\frac{1}{2}}}{|x|}\left(\frac{|x|^\alpha}{(2K)^\alpha t}\right) \\ &\times \sum_{\nu=0}^\infty \frac{(-1)^\nu}{\nu!}\frac{1}{\Gamma\left(\frac{1+\alpha}{2} + \frac{\alpha}{2}\nu\right)\Gamma\left(1 - \frac{\alpha}{2}(1+\nu)\right)}\left(\frac{|x|^\alpha}{(2K)^\alpha t}\right)^\nu, \alpha \neq 1, \alpha \neq 2.\end{aligned}$$

6.4 Discussion

The first solar neutrino experiment led by Raymond Davis Jr. showed a deficit of neutrinos relative to the solar model prediction, referred to as the solar neutrino problem since the 1970s. The Kamiokande experiment led by Masatoshi Koshiba successfully observed solar neutrinos, as first reported in 1980s. The solar neutrino problem was solved due to neutrino oscillations by comparing the SuperKamiokande and Sudbury Neutrino Observatory results. While recent decades have offered tremendous advances in solar neutrinos across the fields of solar physics (Buldgen et al. 2024; Yang and Tian 2024), nuclear physics (Bertulani et al. 2022; Hwang et al. 2023), and neutrino physics (Sturrock et al. 2021; Slad 2024), many lingering mysteries remain.

This chapter takes advantage of publicly available solar neutrino SuperKamiokande data and analyses them by applying diffusion entropy analysis and standard deviation analysis. The result is a scaling exponent $\delta < 1$ indicating anomalous diffusion of solar neutrinos in terms of Lévy flights. Based on this result the chapter developes the probability density function for neutrino flights and derives the respective differential equation in terms of a fractional Fokker-Planck equation. Accordingly, the closed form analytic representation of the neutrino power density function is given as a Fox H-function that can be used for further numerical exercises for the benefit of solar neutrino physics.

The authors have grateful for the support in the diffusion entropy analysis and standard deviation analysis by Dr. Alexander Haubold while he was doing his research at the Department of Computer Science, Columbia University, New York (USA). The results of diffusion entropy analysis were independently confirmed by the research of Dr. Nicy Sebastian, Department of Statistics, St Thomas College, Thrissur, University of Calicut, Kerala (India).

References

Abe, K. et al.: Search for Periodic Time Variations of the Solar ^8B Neutrino Flux Between 1996 and 2018 in SuperKamiokande (2024a). http://arxiv.org/pdf/2311.01159

Abe, K. et al.: Solar neutrino measurements using the full data period of SuperKamiokande-IV. Phys. Rev. D **109** (2024b). https://doi.org/10.1103/PhysRevD.109.092001

Bertulani, C.A., Hall, F.W., Santoyo, B.I.: Big Bang nucleosynthesis as a probe of new physics. arXiv:2210.04071

Buldgen, G., Noels, A., Scuflaire, R., Amarsi, A.M., Grevesse, N., Eggenberger, P., Colgan, J., Fontes, C.J., Baturin, V.A., Oreshina, A.V., Ayukov, S.V., Hakel, P., Kilcrease, D.P.: In-depth analysis of solar models with high-metallicity abundances and updated opacity tables. arXiv:2404.10478

Cravens, J.P., et al.: Solar neutrino measurements in SuperKamiokande-II. Phys. Rev. D **78**, 032002 (2008)

Culbreth, G., Baxley, J., Lambert, D.: Detecting temporal scaling with modified diffusion entropy analysis. arXiv:2311.11453

Haubold, A., Haubold, H.J., Kumar, D.: Heliosheath: Diffusion entropy analysis and nonextensivity q-triplet. arXiv: 1202.3417v1 [physics.gen-ph]

Hilfer, R.: Mathematical and physical interpretation of fractional derivatives and integrals. In: Kochubei, A., Luchko, Yu. (eds.) Handbook of Fractional Calculus with Applications, Volume 1: Basic Theory, pp. 47–85. de Gruyter, Berlin (2018)

Hwang, E., Ko, H., Heo, K., Cheoun, M.-K., Jang, D.: Revisiting the Gamow factor reactions on light nuclei. arXiv:2302.10102

Mathai, A.M., Haubold, H.J.: Erdélyi-Kober Fractional Calculus: From a Statistical Perspective. Inspired by Solar Neutrino Physics. Springer, Singapore (2018)

Mathai, A.M., Saxena, R.K., Haubold, H.J.: The H-Function: Theory and Applications. Springer, New York (2010)

Metzler, R., Klafter, J.: The random walk's guide to anomalous diffusion: a fractional dynamics approach. Phys. Rep. **339**, 1–77 (2000)

Orebi Gann, G.D., Zuber, K., Bemmerer, D., Serenelli, A.: The Future of Solar Neutrinos. Annu. Rev. Nucl. Part. Sci. **71**, 491–528 (2021)

Sakurai, K.: Solar Neutrino Problems - How They Were Solved. TERRAPUB, Tokyo (2018)

Scafetta, N.: Fractal and Diffusion Entropy Analysis of Time Series: Theory, concepts, applications and computer codes for studying fractal noises and Lévy walk signals. VDM Verlag Dr. Mueller, Saarbruecken (2010)

Slad, L.M.: Logic and numbers related to solar neutrinos. Annalen der Physik (Berlin) **536**, 2400168 (2024)

Sturrock, P.A., Piatibratova, O., Scholkmann, F.: Comparative analysis of SuperKamiokande solar neutrino measurements and geological survey of Israel radon decay measurements. Front. Phys. **9**, 718306. https://doi.org/10.3389/fphy.2021.718306

Tsallis, C.: Reminiscenses of half a century of life in the world of theoretical physics. Entropy **26**, 158 (2024). https://doi.org/10.3390/e26020158

Yang, W., Tian, Z.: Solar models and astrophysical S-factors constrained by helioseismic results and updated neutrino fluxes. https://arxiv.org/abs/2405.10472

Yoo, J., et al.: Search for periodic modulations of the solar neutrino flux in SuperKamiokande-I. Phys. Rev. D **68**, 092002 (2003)

Open Access This chapter is licensed under the terms of the Creative Commons Attribution 4.0 International License (http://creativecommons.org/licenses/by/4.0/), which permits use, sharing, adaptation, distribution and reproduction in any medium or format, as long as you give appropriate credit to the original author(s) and the source, provide a link to the Creative Commons license and indicate if changes were made.

The images or other third party material in this chapter are included in the chapter's Creative Commons license, unless indicated otherwise in a credit line to the material. If material is not included in the chapter's Creative Commons license and your intended use is not permitted by statutory regulation or exceeds the permitted use, you will need to obtain permission directly from the copyright holder.

Chapter 7
Neutrino Astrophysics, 2025 Update: Neutrino Masses and Mixings

7.1 Introduction

The Standard Model (SM) of Particle Physics is the pinnacle of the understanding of neutrino physics (Deppisch 2019; Oberauer et al. 2020). It comes with a plethora of parameters, the masses and the flavour mixings, that are seemingly not fixed by any known fundamental principle. In the SM, the neutrino spectrum is simple: all neutrinos are massless. Neutrino oscillations, where neutrinos seemingly change flavour in flight, cannot be accommodated in the SM due to the mass of the neutrinos. Neutrino oscillations thus imply massive neutrino eigenstates, and the SM must be extended. Moreover, neutrino oscillation experimental data suggest that the neutrino spectrum is not hierarchical, with three massive light neutrinos and a mixing matrix exhibiting near-maximal mixing (Deppisch 2019; Oberauer et al. 2020).

To make sense of the neutrino sector, it was argued that the light neutrino mass matrix could be generated randomly from a more fundamental Dirac neutrino mass matrix and a more fundamental Majorana neutrino mass matrix with random elements distributed according to a Gaussian ensemble, a principle dubbed the anarchy principle (Haba and Murayama 2001). These more fundamental neutrino mass matrices would come from the extended SM where the seesaw mechanism occurs (Yanagida 1979). It was argued that the probability density function (PDF) for the mixing angles and phases is the appropriate Haar measure of the symmetry group, implying near-maximal mixings (Fortin et al. 2016; Haba et al. 2023; Fortin et al. 2020). We have shown that the PDF can be obtained either by analysing data by diffusion entropy analysis (done for the case of solar neutrinos from observations emanating from SuperKamiokande) (Scafetta 2010; Mathai and Haubold 2018; Haubold and Mathai 2024) or as proceeding in this chapter. Then, the anarchy principle was analysed mostly numerically, reaching conclusions, for example about the preferred normal hierarchy of the neutrino masses.

Although several numerical results have been obtained, few analytical results on the seesaw ensemble, which is derived from the anarchy principle, exist. That it is the case even though random matrix theory is a well-studied subject in mathematics is surprising. It is therefore clear that a thorough analytical investigation of the seesaw ensemble is possible.

It is an open issue to investigate analytically the seesaw ensemble derived from the anarchy principle with the help of the usual tools of random matrix theory. The seesaw ensemble PDF can be obtained from $N \times N$ fundamental Dirac and Majorana neutrino mass matrices with real or complex elements. The joint PDF for the singular (eigen) values in the complex (real) case can be derived and it can be shown that the group variables decouple straightforwardly as in the usual Gaussian ensembles.

The following notations will be used in this chapter: Real scalar (1×1 matrix) variables, whether random or mathematical, will be denoted by lower-case letters such as x, y, z. Real vector ($1 \times p$ or $p \times 1$, $p > 1$, matrix), matrix ($p \times q$) variables will be denoted by capital letters such as X, Y. Variables in the complex domain will be indicated with a tilde such as $\tilde{x}, \tilde{y}, \tilde{X}, \tilde{Y}$. Constants will be denoted by a, b etc. for scalars and A, B etc. for vectors/matrices. No tilde will be used on constants. For a $p \times p$ matrix B, $|B|$ or $\det(B)$ will denote the determinant of the matrix B. When B is in the complex domain, one can write $|B| = a + ib, i = \sqrt{(-1)}, a, b$ are real scalars. Then, the absolute value of the determinant will be denoted by $|\det(B)| = |\det(BB^*)|^{\frac{1}{2}} = \sqrt{(a^2 + b^2)}$ where B^* means the conjugate transpose of B, that is $B^* = (B')^c = (B^c)'$ where a prime denotes the transpose and c in the exponent denotes the conjugate. If $X = (x_{jk})$ is a $p \times q$ matrix in the real domain, where the x_{jk}'s are distinct real scalar variables, then the wedge product of their differentials will be denoted as $dX = \wedge_{j=1}^{p} \wedge_{k=1}^{q} dx_{jk}$. If $X = X'$ (real symmetric), then $dX = \wedge_{j \geq k} dx_{jk}$. For two real scalar variables x_1 and x_2, the wedge product is defined as $dx_1 \wedge dx_2 = -dx_2 \wedge dx_1$ so that $dx_j \wedge dx_j = 0, j = 1, 2$. Also, $\int_X f(X) dX$ will denote the real-valued scalar function $f(X)$ of X is integrated over X. If the $p \times q$ matrix $\tilde{X}$ is in the complex domain, then $\tilde{X} = X_1 + iX_2, i = \sqrt{(-1)}, X_1, X_2$ are real $p \times q$ matrices, then $d\tilde{X} = dX_1 \wedge dX_2$. Other notations will be explained whenever they occur for the first time.

This chapter is organized as follows: Sect. 7.2 gives a mathematical introduction to the models. Section 7.3 derives the distribution of the light neutrino mass matrix in explicit computable form. Section 7.4 is providing the densities in terms of the eigenvalues, including the cases of the largest eigenvalue and the smallest eigenvalue.

7.2 Modified Dirac and Majorana Neutrino Matrices and Their Distributions

Let X and $\tilde{X}$ be $p \times n$, $p \leq n$ matrix of rank p in the real and complex domains respectively. If X and $\tilde{X}$ are $p \times n$ matrix-variate random variables in the real and complex domains respectively, then this matrix X corresponds to the $N \times N$ Dirac matrix M_D considered in Fortin et al. (2016); Haba et al. (2023); Fortin et al. (2020). Let $Y > O, \tilde{Y} > O$ be $n \times n$ real positive definite and Hermitian positive definite matrices in the real and complex domains respectively, where $(\cdot) > O$ denotes the

7.2 Modified Dirac and Majorana Neutrino Matrices and Their Distributions

matrix ($\cdot$) is real positive definite or Hermitian positive definite. If Y and $\tilde{Y}$ are $n \times n$ matrix-variate random variables in the real and complex domains, then this Y corresponds to the $N \times N$ Majorana matrix M_R in Fortin et al. (2016); Haba et al. (2023); Fortin et al. (2020). Let $U = -XY^{-1}X$ where U corresponds to the light neutrino mass matrix M_ν in Fortin et al. (2016); Haba et al. (2023); Fortin et al. (2020). Due to our assumption of X being a full rank matrix, the rows of X are linearly independent so that a singular distribution for any column of X is avoided. The columns of X are $p \times 1$ which corresponds to a p-vector in multivariate statistical analysis. Also if the columns of X are iid (independently and identically distributed) then X can represent a sample matrix of a sample of size n from a p-variate population. When $p = n$, X will be a $n \times n$ square matrix as considered in Fortin et al. (2016); Haba et al. (2023); Fortin et al. (2020). Let X have a $p \times n$ matrix-variate Gaussian density in the standard form, denoted by $f_1(X)$, where

$$f_1(X)\mathrm{d}X = c_1 e^{-\mathrm{tr}(XX')} \mathrm{d}X, \quad f_1(\tilde{X})\mathrm{d}\tilde{X} = \tilde{c}_1 e^{-\mathrm{tr}(\tilde{X}\tilde{X}^*)} \mathrm{d}\tilde{X} \qquad (7.1)$$

where c_1 is the normalizing constant, $c_1 = \pi^{pn/2}$. If a real scalar scaling constant $b_1 > 0$ in introduced in the exponent in (7.1) and write the exponent as $-b_1 \mathrm{tr}(XX')$, then the normalizing constant changes to $(b_1\pi)^{pn/2}$. In the corresponding complex case, $\tilde{c}_1 = \pi^{pq}$ and if a real scaling factor $b_1 > 0$ is present, then $\tilde{c}_1 = (b_1\pi)^{pq}$ respectively. If a location parameter matrix is to be included, then replace X in (2.1) by $X - M$ and $\tilde{X}$ by $\tilde{X} - \tilde{M}$ where the M and $\tilde{M}$ matrices can act as the mean value or expected value of X and expected value of $\tilde{X}$ respectively, where $M = E[X]$, $\tilde{M} = E[\tilde{X}]$ where $E[(\cdot)]$ denotes the expected value of ($\cdot$). There is not going to be any change in the normalizing constants c_1 and $\tilde{c}_1$. If scaling matrices are also to be inserted, then replace XX' by $A(X - M)B(X - M)'$ and $\tilde{X}\tilde{X}^*$ by $A(\tilde{X} - \tilde{M})B(\tilde{X} - \tilde{M})^*$ respectively, where $A > O$ is $p \times p$ and $B > O$ is $n \times n$ real positive definite constant matrices in the real case, and $A = A^* > O$ and $B = B^* > O$ (Hermitian positive definite matrices) in the complex domain. The normalizing constants will change to $c_1 = \pi^{pq/2}|A|^{\frac{n}{2}}|B|^{\frac{p}{2}}$ and $\tilde{c}_1$ changes to $\tilde{c}_1 = \pi^{pq}|\det(A)|^n|\det(B)|^p$ respectively. These changes are taking place due to Lemma 2.1 given below. When XX' is changed to $AXBX'$, one can give interpretations in terms of the covariance matrices of the columns and rows of X. For example, the inverse of A can act as the covariance matrix of each $p \times 1$ column vector of X and similarly the inverse of B can act as the covariance matrix of each row of X when X is a sample matrix from some p-variate population. The above are some of the advantages of considering the $p \times n$ matrix X and inserting location parameter matrix and scaling matrices in XX'. Corresponding interpretations can be given in the complex case also.

Lemma 7.1 *Let $X = (x_{jk})$ be a $p \times q$ matrix in the real domain with pq distinct real scalar variables as elements x_{jk}'s. Let A be a $p \times p$ and B be a $q \times q$ real nonsingular constant matrices. Consider the linear transformation $Y = AXB$. Then,*

$$Y = AXB, |A| \neq 0, |B| \neq 0, \Rightarrow \mathrm{d}Y = |A|^q |B|^p \mathrm{d}X$$

In the corresponding complex domain, $\tilde{Y} = A\tilde{X}B$ where A and B are nonsingular and they may be in the real or complex domain, then

$$\tilde{Y} = A\tilde{X}B, |A| \neq 0, |B| \neq 0, \Rightarrow d\tilde{Y} = |\det(A)|^{2q}|\det(B)|^{2p}d\tilde{X}.$$

In the discussion to follow, we will need the Jacobian from a symmetric transformation, for example, $p = q$, $B = A'$ in the real case in Lemma 7.1. In the symmetric case, the Jacobian will not be available from Lemma 7.1.

Lemma 7.2 *Consider the $p \times p$ matrices $X = X'$, A and $\tilde{X} = \tilde{X}^*$, A in the real and complex domains respectively where A is a nonsingular constant matrix and in the complex case A could be real or complex. Note that we assume X is symmetric in the real domain and $\tilde{X}$ is Hermitian in the complex domain. Then,*

$$Y = AXA', |A| \neq 0, \Rightarrow dY = |A|^{p+1}dX, \tilde{Y} = A\tilde{X}A^* \Rightarrow d\tilde{Y} = |\det(A)|^{2p}d\tilde{X}.$$

If X is skew symmetric then the exponent $p + 1$ changes to $p - 1$ in the Jacobian part in Lemma 7.2. If $\tilde{X}$ is skew Hermitian, there is no change, $2p$ in the exponent will remain as $2p$.

Let the $n \times n$ matrix Y, corresponding to the Majorana neutrino mass matrix M_R in Fortin et al. (2016); Haba et al. (2023); Fortin et al. (2020) have the following density:

$$f_2(Y)dY = c_2|Y|^{\alpha - \frac{n+1}{2}}e^{-\text{tr}(Y)}dY, \quad f_2(\tilde{Y})d\tilde{Y} = \tilde{c}_2|\det(\tilde{Y})|^{\alpha - n}e^{-\text{tr}\tilde{Y}}d\tilde{Y} \quad (7.2)$$

where $Y = Y' > O$ (real positive definite), $\tilde{Y} = \tilde{Y}^* > O$ (Hermitian positive definite). Here α is a free parameter which may be given some physical interpretations. The $n \times n$ positive definite matrix Y can always be written as $Y_1 Y_1'$ where Y_1 is a $n \times n_1, n \leq n_1$ matrix of rank n where n_1 can be equal to n or Y_1 can be a square matrix also. Further, Y_1 can be a square root of Y where a square root can be uniquely defined when Y is positive definite or $\tilde{Y}$ is Hermitian positive definite. Thus, the exponents in the densities of Y and $\tilde{Y}$ can have exactly the same structures as in the corresponding densities in Fortin et al. (2016); Haba et al. (2023); Fortin et al. (2020). But in Fortin et al. (2016); Haba et al. (2023); Fortin et al. (2020) the densities are Gaussian forms and in (7.2) above, the densities are real and complex matrix-variate Gamma densities for Y and $\tilde{Y}$. The normalizing constants c_2 and $\tilde{c}_2$ are the following:

$$c_2 = \frac{1}{\Gamma_n(\alpha)}, \Re(\alpha) > \frac{n-1}{2}, \tilde{c}_2 = \frac{1}{\tilde{\Gamma}_n(\alpha)}, \Re(\alpha) > n - 1$$

where $\Re(\cdot)$ means the real part of $(\cdot)$ and $\Gamma_n(\alpha)$ and $\tilde{\Gamma}_n(\alpha)$ are the real and complex matrix-variate gamma functions defined as the following:

$$\Gamma_n(\alpha) = \begin{cases} \pi^{\frac{n(n-1)}{4}}\Gamma(\alpha)\Gamma(\alpha - \frac{1}{2})\cdots\Gamma(\alpha - \frac{n-1}{2}), \Re(\alpha) > \frac{n-1}{2} \\ \int_{Z>O} |Z|^{\alpha - \frac{n+1}{2}}e^{-\text{tr}(Z)}dZ \end{cases}$$

7.3 Derivation of the Density of U

$$\tilde{\Gamma}_n(\alpha) = \begin{cases} \pi^{\frac{n(n-1)}{2}} \Gamma(\alpha)\Gamma(\alpha-1)\cdots\Gamma(\alpha-n+1), \Re(\alpha) > n-1 \\ \int_{\tilde{Z}>O} |\det(\tilde{Z})|^{\alpha-n} e^{-\operatorname{tr}(\tilde{Z})} \mathrm{d}\tilde{Z}, \Re(\alpha) > n-1. \end{cases} \quad (7.3)$$

Let $Y^{\frac{1}{2}}$ be the positive definite square root of the positive definite matrix $Y > O$. Then, if we consider X of (7.1) to be scaled by $XY^{-\frac{1}{2}}$ then this scaling has the effect of making the rows of X correlation-free if Y is the correlation matrix of each row of X. Thus, in a physical situation if the rows are likely to be correlated then they can be made correlation free by scaling with the proper scaling matrix, namely the square root of the inverse of the correlation matrix. In the scaled XX', namely $AXBX'$ considered above, B^{-1} corresponds to Y in the present discussion. When scaled with the proper scaling matrix XX' goes to $(XY^{-\frac{1}{2}})(XY^{-\frac{1}{2}})' = XY^{-1}X'$ and one has similar changes in the complex case also. Hence, our light neutrino mass matrix $U = XY^{-1}X$ has proper interpretations in terms of scaling models, making rows correlation free etc.

Our interest is to derive the density of U. For this purpose, we need either the assumption that X and Y are independently distributed, in that case the joint density of X and Y is $f_1(X) f_2(Y)$, the product, or we have to assume that $f_1(X)$ is a conditional density, in the sense, for every given Y, one has the density of X a matrix-variate Gaussian as in (7.1) and $f_2(Y)$ is then the marginal density of Y and again the joint density will be the product $f_1(X) f_2(Y)$. We will assume $f_1(X)$ being a conditional density of X for every given Y and derive the density of $U = XY^{-1}X'$ and $\tilde{U} = \tilde{X}\tilde{Y}^{-1}\tilde{X}^*$.

7.3 Derivation of the Density of U

In our notation, the light neutrino mass matrix is $U = -XY^{-1}X'$ in the real case and $\tilde{U} = -\tilde{X}\tilde{Y}^{-1}\tilde{X}^*$ in the complex case. Let us consider the real case first. Ignoring the sign, U can be written as $U = (XY^{-\frac{1}{2}})(XY^{-\frac{1}{2}})'$ since Y is symmetric real positive definite. Let $Z = XY^{-\frac{1}{2}} \Rightarrow X = ZY^{\frac{1}{2}}$ and $\mathrm{d}X = |Y|^{\frac{p}{2}}\mathrm{d}Z$, for fixed Y, by using Lemma 7.1, and $U = ZZ'$. The joint density of X and Y, is the conditional density of X, given Y, times the marginal density of Y. That is, denoting the joint density by $f(X, Y)$, we have the following:

$$f(X, Y)\mathrm{d}X \wedge \mathrm{d}Y = c_1 c_2 |Y|^{\alpha - \frac{n+1}{2}} e^{-\operatorname{tr}(ZYZ')} e^{-\operatorname{tr}(Y)} |Y|^{\frac{p}{2}} \mathrm{d}Z \wedge \mathrm{d}Y$$
$$= c_1 c_2 |Y|^{\alpha + \frac{p}{2} - \frac{n+1}{2}} e^{-\operatorname{tr}[Y(I_n + Z'Z)]} \mathrm{d}Z \wedge \mathrm{d}Y.$$

But

$$\operatorname{tr}(Y + ZYZ') = \operatorname{tr}(Y(I_n + Z'Z)) = \operatorname{tr}[(I_n + Z'Z)^{\frac{1}{2}} Y (I_n + Z'Z)^{\frac{1}{2}}]$$

Even though $Z'Z$ is singular and positive semi-definite, due to the presence of the identity matrix $I = I_n$, we may take $I_n + Z'Z$ to be positive definite and hence one may consider the positive definite square root of $I_n + Z'Z$. Now, we can integrate out Y by using a real matrix-variate gamma of (7.3). That is,

$$\int_{Y>O} |Y|^{\frac{p}{2}+\alpha-\frac{n+1}{2}} e^{-\text{tr}[(I_n+Z'Z)^{\frac{1}{2}}Y(I_n+Z'Z)^{\frac{1}{2}}]} dY = \Gamma_n(\alpha + \frac{p}{2})|I_n + Z'Z|^{-(\alpha+\frac{p}{2})}, \Re(\alpha) > \frac{n-1}{2}.$$

But we can write $|I_n + Z'Z|$ in terms of the $p \times p$ real positive definite matrix ZZ'. Consider the expansion of the following determinant in two different ways in terms of its submatrices, denoting the determinant by η:

$$\eta = \begin{vmatrix} I_p & -Z \\ Z' & I_n \end{vmatrix} = |I_p| |I_n - Z'I_p^{-1}(-Z)| = |I_p| |I_n + Z'Z| = |I_n + Z'Z|$$

$$\eta = |I_n| |I_p - (-Z)I_n^{-1}Z'| = |I_n| |I_p + ZZ'| = |I_p + ZZ'| \Rightarrow |I_n + Z'Z| = |I_p + ZZ'|. \quad (7.4)$$

Hence, the density of Z, denoted by $g(Z)$, is the following:

$$g(Z)dZ = c_1 c_2 \Gamma_n(\alpha + \frac{p}{2})|I_p + ZZ'|^{-(\alpha+\frac{p}{2})} dZ. \quad (7.5)$$

Going through steps parallel to the real case, one has the corresponding result in the complex case, denoted by $\tilde{g}(\tilde{Z})$ as the following:

$$\tilde{g}(\tilde{Z})d\tilde{Z} = \tilde{c}_1 \tilde{c}_2 \tilde{\Gamma}_n(\alpha + p)|\det(I + \tilde{Z}\tilde{Z}^*)|^{-(\alpha+p)} d\tilde{Z}. \quad (7.6)$$

Our matrix is $U = ZZ'$. We can go from the density of Z to the density of $U = ZZ'$ by using the following result from Mathai (1997) which will be stated as a lemma. This result is available in Chap. 5 as Lemma 5.1. For the sake of ready reference, this result is given as the next lemma.

Lemma 7.2 *Let $X = (x_{jk})$ be a $p \times q$, $p \leq q$ matrix of rank p in the real domain where the x_{jk}'s are distinct real scalar variables. Let $S = XX'$. Then, going through a transformation involving a lower triangular matrix with positive diagonal elements and a unique semi-orthonormal matrix and then integrating out the differential element corresponding to the semi-orthonormal matrix, we have the following relationship between dX and dS:*

$$dX = \frac{\pi^{\frac{pq}{2}}}{\Gamma_p(\frac{q}{2})} |S|^{\frac{q}{2}-\frac{p+1}{2}} dS.$$

In the corresponding complex case, let $\tilde{X}$ be $p \times q$, $p \leq q$ matrix in the complex domain with distinct scalar complex variables as elements. Let $\tilde{S} = \tilde{X}\tilde{X}^*$. Then, going through a transformation involving a lower triangular matrix with real and positive diagonal elements and a unique semi-unitary matrix and then integrating out the differential element corresponding to the semi-unitary matrix, we have the following connection:

7.4 Densities in Terms of the Eigenvalues

$$d\tilde{X} = \frac{\pi^{pq}}{\tilde{\Gamma}_p(q)} |\det(\tilde{S})|^{q-p} d\tilde{S}.$$

With the help of Lemma 7.2, we can go to the density of Z in (7.5) to the density of $U = ZZ'$, denoted by $g_1(U)$. Since the variable is changed from a $p \times n$ matrix to a $p \times p$ matrix, the normalizing constant will change. Hence we may write

$$g_1(U) dU = c|U|^{\frac{n}{2} - \frac{p+1}{2}} |I + U|^{-(\alpha + \frac{p}{2})} dU \qquad (7.7)$$

for $\Re(\alpha) > \frac{n-1}{2}$, $n > p - 1$, where c is the corresponding normalizing constant. This $g_1(U)$ is a real matrix-variate type 2 beta density with the parameters $(\frac{n}{2}, \alpha + \frac{p-n}{2})$. Hence, the normalizing constants, denoted by c in the real case and $\tilde{c}$ in the complex case, are the following:

$$c = \frac{\Gamma_p(\frac{n}{2})\Gamma_p(\alpha + \frac{p}{2} - \frac{n}{2})}{\Gamma_p(\alpha + \frac{p}{2})}, \Re(\alpha) > \frac{n-1}{2}, \tilde{c} = \frac{\tilde{\Gamma}_p(n)\tilde{\Gamma}_p(\alpha + p - n)}{\tilde{\Gamma}_p(\alpha + p)}, \Re(\alpha) > n - 1, \qquad (7.8)$$

evaluated from real and complex $p \times p$ matrix-variate type 2 beta densities respectively. Using steps parallel to the real case, we have the corresponding density $\tilde{g}_1(\tilde{U})$ in the complex case as the following, for $\Re(\alpha) > n - 1$:

$$\tilde{g}_1(\tilde{U}) d\tilde{U} = \tilde{c}|\det(\tilde{U})|^{\alpha - p} |\det(I + \tilde{U})|^{-(\alpha + n)} d\tilde{U} \qquad (7.9)$$

where $\tilde{c}$ is given in (7.8).

7.4 Densities in Terms of the Eigenvalues

From (7.7) the $p \times p$ real positive definite matrix U has a real matrix-variate type 2 beta density with the parameters $(\frac{n}{2}, \alpha + \frac{p}{2} - \frac{n}{2})$ with $\Re(\alpha) > \frac{n}{2} - \frac{p}{2} + \frac{p-1}{2} = \frac{n-1}{2}$, that is, with the density

$$g_1(U) dU = \frac{\Gamma_p(\alpha + \frac{p}{2})}{\Gamma_p(\frac{n}{2})\Gamma_p(\alpha + \frac{p}{2} - \frac{n}{2})} |U|^{\frac{n}{2} - \frac{p+1}{2}} |I + U|^{-(\alpha + \frac{p}{2})} dU \qquad (7.10)$$

for $\Re(\alpha) > \frac{n-1}{2}$. The corresponding density in the complex domain is the following where $\tilde{U}$ is Hermitian positive definite and $|\det(\tilde{U})|$ means the absolute value of the determinant of $\tilde{U}$:

$$\tilde{g}_1(\tilde{U}) d\tilde{U} = \frac{\tilde{\Gamma}_p(\alpha + p)}{\tilde{\Gamma}_p(n)\tilde{\Gamma}_p(\alpha + p - n)} |\det(\tilde{U})|^{n-p} |\det(I + \tilde{U})|^{-(\alpha + p)} d\tilde{U} \qquad (7.11)$$

for $\Re(\alpha) > n - 1$. We can convert U and $\tilde{U}$ and write the densities in terms of their eigenvalues. If μ_j is an eigenvalue of U, then $0 < \mu_j < \infty$, $j = 1, \ldots, p$. Similar is the case for the eigenvalues of $\tilde{U}$. For convenience, let us convert U and $\tilde{U}$ to the corresponding type 1 beta form. Consider the transformation

$$V = (I + U)^{-\frac{1}{2}} U (I + U)^{-\frac{1}{2}}, \ \tilde{V} = (I + \tilde{U})^{-\frac{1}{2}} \tilde{U} (I + \tilde{U})^{-\frac{1}{2}}.$$

Then, V and $\tilde{V}$ will be $p \times p$ matrix-variate type 1 beta with the same parameters, see Mathai (1997); Mathai et al. (2022). Let the densities of V and $\tilde{V}$ be denoted by $g_2(V)$ and $\tilde{g}_2(\tilde{V})$ respectively. Then,

$$g_2(V)dV = \frac{\Gamma_p(\alpha + \frac{p}{2})}{\Gamma_p(\frac{n}{2})\Gamma_p(\alpha + \frac{p}{2} - \frac{n}{2})} |V|^{\frac{n}{2} - \frac{p+1}{2}} |I - V|^{\alpha + \frac{p}{2} - \frac{n}{2} - \frac{p+1}{2}} dV \quad (7.12)$$

and

$$\tilde{g}_2(\tilde{V})d\tilde{V} = \frac{\tilde{\Gamma}_p(\alpha + p)}{\tilde{\Gamma}_p(n)\tilde{\Gamma}_p(\alpha + p - n)} |\det(\tilde{V})|^{n-p} |\det(I - \tilde{V})|^{\alpha - n} d\tilde{V} \quad (7.13)$$

for $\Re(\alpha) > \frac{n-1}{2}, n-1$ respectively in the real and the corresponding complex case. If λ_j is an eigenvalue of V, then $\lambda_j = \frac{\mu_j}{(1+\mu_j)} \Rightarrow \mu_j = \frac{\lambda_j}{(1-\lambda_j)}, 0 < \lambda_j < 1, 0 < \mu_j < \infty, j = 1, \ldots, p$. Let Q be a $p \times p$ unique orthonormal matrix, $QQ' = I$, $Q'Q = I$ such that $Q'VQ = \text{diag}(\lambda_1, \ldots, \lambda_p)$ with $1 > \lambda_1 > \lambda_2 > \cdots > \lambda_p > 0$. Correspondingly, let $\tilde{Q}$ be a unique unitary matrix, $\tilde{Q}\tilde{Q}^* = I$, $\tilde{Q}^*\tilde{Q} = I$ such that $\tilde{Q}^*\tilde{V}\tilde{Q} = \text{diag}(\lambda_1, \ldots, \lambda_p)$, where $\tilde{Q}^*$ means the conjugate transpose of $\tilde{Q}$. When λ_j's are real scalar variables we can assume $Pr\{\lambda_i = \lambda_j, i \neq j\} = 0$ almost surely. Hence, without loss of generality we assume that the λ_j's are distinct, $1 > \lambda_1 > \cdots > \lambda_p > 0$. Observe that the eigenvalues of Hermitian matrices are also real and hence the eigenvalues of both V and $\tilde{V}$ will be real and we will denote them by the same symbols λ_j's. Also, $Q'VQ = D = \text{diag}(\lambda_1, \ldots, \lambda_p) \Rightarrow V = QDQ'$, $|V| = \lambda_1 \cdots \lambda_p$, $|I - V| = \prod_{j=1}^{p}(1 - \lambda_j)$ and when V is transformed to its eigenvalues a factor $\prod_{i<j}(\lambda_i - \lambda_j)$ comes in, both in the real and complex cases, see Mathai (1997); Mathai et al. (2022). If the differential elements corresponding to Q and $\tilde{Q}$ are denoted by dG and $d\tilde{G}$ respectively, then from Mathai (1997); Mathai et al. (2022), $G = Q'(dQ)$, $\tilde{G} = \tilde{Q}(d\tilde{Q})$ where, for example, (dQ) is the matrix of differentials in Q and the integrals over dG and $d\tilde{G}$ are the following results which will be written as a lemma, see Mathai (1997); Mathai et al. (2022):

Lemma 7.3 *For the $G, dG, \tilde{G}, d\tilde{G}$ as defined above, we have*

$$\int dG = \frac{\pi^{\frac{p^2}{2}}}{\Gamma_p(\frac{p+1}{2})}, \int d\tilde{G} = \frac{\pi^{p(p-1)}}{\tilde{\Gamma}_p(p)}.$$

7.4 Densities in Terms of the Eigenvalues

Let us verify this lemma for $p = 2, 3$. For a $p \times p$ real positive definite matrix X we have

$$\int_{X>O} |X|^{\alpha - \frac{p+1}{2}} e^{-\text{tr}(X)} dX = \Gamma_p(\alpha), \ \Re(\alpha) > \frac{p-1}{2}$$

from the real matrix-variate gamma integral. In the complex case, let the $p \times p$ matrix $\tilde{X}$ be Hermitian positive definite. Then, from the complex matrix-variate gamma integral we have

$$\int_{\tilde{X}>O} |\det(\tilde{X})|^{\alpha - p} e^{-\text{tr}(\tilde{X})} d\tilde{X} = \tilde{\Gamma}_p(\alpha), \ \Re(\alpha) > p - 1.$$

Consider the integrals in the real and complex cases when $\alpha = \frac{p+1}{2}$ in the real case and $\alpha = p$ in the complex case. Then,

$$\int_{X>O} e^{-\text{tr}(X)} dX = \Gamma_p(\frac{p+1}{2}), \ \int_{\tilde{X}>O} e^{-\text{tr}(\tilde{X})} d\tilde{X} = \tilde{\Gamma}_p(p).$$

If we go through a unique orthonormal transformation involving an orthonormal matrix Q then in the real case

$$\Gamma_p(\frac{p+1}{2}) = \int_D \{\prod_{i<j}(\lambda_i - \lambda_j)\} e^{-\text{tr}(D)} dD \int_Q dG, \ \int_Q dG = \frac{\pi^{\frac{p^2}{2}}}{\Gamma_p(\frac{p}{2})}$$

and in the corresponding complex case

$$\tilde{\Gamma}_p(p) = \int_D \{\prod_{i<j}(\lambda_i - \lambda_j)^2\} e^{-\text{tr}(D)} dD \int_{\tilde{Q}} d\tilde{G} = \frac{\pi^{p(p-1)}}{\tilde{\Gamma}_p(p)}.$$

Then, in the real case, for $p = 2$, $\Gamma_p(\frac{p+1}{2}) = \Gamma_2(\frac{3}{2}) = \pi/2$. $\frac{\pi^{\frac{p^2}{2}}}{\Gamma_p(\frac{p}{2})} = \frac{\pi^2}{\Gamma_2(1)} = \pi$. Now, $\Gamma_p(\frac{p+1}{2})$ divided by the right side quantity π gives $\frac{1}{2}$. Now, consider the integral over D for $p = 2$ in the real case. Let $u_1 = \lambda_1 - \lambda_2$.

$$\int_D (\lambda_1 - \lambda_2) e^{-(\lambda_1 + \lambda_2)} dD = \int_{u_1=0}^{\infty} u_1 e^{-u_1} du_1 \int_{\lambda_2=0}^{\infty} e^{-2\lambda_2} d\lambda_2$$
$$= \frac{1}{2}.$$

Hence, for $p = 2$ in the real case, Lemma 4.1 is verified. Now, for $p = 3$ in the real case, the left side quantity $\Gamma_p(\frac{p+1}{2}) = \Gamma_3(2) = \frac{\pi^2}{2}$. $\frac{\pi^{\frac{p^2}{2}}}{\Gamma_p(\frac{p}{2})} = \frac{\pi^{9/2}}{\Gamma_3(\frac{3}{2})} = 2\pi^2$. Then, $\Gamma_p(\frac{p+1}{2})/[\pi^{\frac{p^2}{2}}/\Gamma_p(\frac{p}{2})] = \frac{\pi^2}{2}/(2\pi^2) = \frac{1}{4}$. Now, consider the integral over D. Let $u_1 = \lambda_1 - \lambda_2, u_2 = \lambda_2 - \lambda_3, u_3 = \lambda_3$. Then,

$$\int_D (\lambda_1 - \lambda_2)(\lambda_1 - \lambda_3)(\lambda_2 - \lambda_3) e^{-(\lambda_1+\lambda_2+\lambda_3)} dD$$
$$= \int_0^\infty \int_0^\infty \int_0^\infty u_1(u_1+u_2)u_2 e^{-(u_1+2u_2+3u_3)} du_1 \wedge du_2 \wedge du_3$$
$$= \frac{1}{3}[\frac{2}{4} + \frac{2}{8}] = \frac{1}{4}.$$

Hence, the result for $p = 3$ in the real case is verified. Now, consider the complex case. The left side quantity is $\tilde{\Gamma}_p(p) = \tilde{\Gamma}_2(2) = \pi^{\frac{p(p-1)}{2}} \Gamma(2)\Gamma(1) = \pi$ for $p = 2$ and $2\pi^3$ for $p = 3$. The right side quantity $[\pi^{p(p-1)}/\tilde{\Gamma}_p(p)] = \pi$ for $p = 2$ and $\frac{\pi^3}{2}$ for $p = 3$. Now, the left side quantity divided by the right side quantity gives the following: $\pi/\pi = 1$ for $p = 2$ and $(2\pi^3)/(\frac{\pi^3}{2}) = 4$ for $p = 3$. Now, consider the integral over D in the complex case. As before, let $u_1 = \lambda_1 - \lambda_2, u_2 = \lambda_2 - \lambda_3$. Then, for $p = 2$,

$$\int_D (\lambda_1 - \lambda_2)^2 e^{-(u_1+2\lambda_2)} du_1 \wedge d\lambda_2$$
$$= \int_0^\infty \int_0^\infty u_1^2 e^{-(u_1+2\lambda_2)} du_1 \wedge d\lambda_2 = 1.$$

Thus, the result for $p = 2$ is verified. Now, consider $p = 3$.

$$\int_0^\infty \int_0^\infty \int_0^\infty u_1^2(u_1+u_2)^2 u_2^2 e^{-(u_1+2u_2+3\lambda_3)} du_1 \wedge du_2 \wedge d\lambda_3$$
$$= \int_0^\infty e^{-3\lambda_3} d\lambda_3 [\int_0^\infty \int_0^\infty (u_1^4 u_2^2 + 2u_1^3 u_2^3 + u_1^2 u_2^4) e^{-u_1-2u_2} du_1 \wedge du_2$$
$$= 4.$$

Hence, the result for $p = 3$ in the complex case is verified.

The joint density of $\lambda_1, \ldots, \lambda_p$ is the following, denoted by $g_3(D)$ in the real case and $\tilde{g}_3(D)$ in the complex case:

$$g_3(D)dD = \frac{\Gamma_p(\alpha + \frac{p}{2})}{\Gamma_p(\frac{n}{2})\Gamma_p(\alpha + \frac{p}{2} - \frac{n}{2})} \frac{\pi^{\frac{p^2}{2}}}{\Gamma_p(\frac{p}{2})} \left\{ \prod_{i<j}(\lambda_i - \lambda_j) \right\}$$
$$\times \left\{ \prod_{j=1}^p \lambda_j^{\frac{n}{2} - \frac{p+1}{2}} \right\} \left\{ \prod_{j=1}^p (1-\lambda_j)^{\alpha - \frac{n-1}{2}} \right\} dD \quad (7.14)$$

$$\tilde{g}_3(D)dD = \frac{\tilde{\Gamma}_p(\alpha + p)}{\tilde{\Gamma}_p(n)\tilde{\Gamma}_p(\alpha + p - n)} \frac{\pi^{p(p-1)}}{\tilde{\Gamma}_p(p)} \{\prod_{i<j}(\lambda_i - \lambda_j)^2\}$$
$$\times \prod_{j=1}^p \lambda_j^{n-p} \left\{ \prod_{j=1}^p (1-\lambda_j)^{\alpha-n} \right\} dD. \quad (7.15)$$

7.4 Densities in Terms of the Eigenvalues

We can write $\prod_{i<j}(\lambda_i - \lambda_j)$ in the real case and $\prod_{i<j}(\lambda_i - \lambda_j)^2$ in the complex case, in the $p \times p$ matrix case, in terms of Vandermonde's determinant.

$$\prod_{i<j}(\lambda_i - \lambda_j) = \begin{vmatrix} \lambda_1^{p-1} & \lambda_1^{p-2} & \cdots & \lambda_1 & 1 \\ \lambda_2^{p-1} & \lambda_2^{p-2} & \cdots & \lambda_2 & 1 \\ \vdots & \vdots & \cdots & \vdots & \vdots \\ \lambda_p^{p-1} & \lambda_p^{p-2} & \cdots & \lambda_p & 1 \end{vmatrix} = A = (a_{ij}), a_{ij} = \lambda_i^{p-j}$$

for all i and j. Let us use the general expansion for a determinant. Then, for $K = (k_1, \ldots, k_p)$ where $k_1, \ldots, k_p$ is a given permutation of $1, 2, \ldots, p$, we have the following:

$$|A| = \sum_K (-1)^{\rho_K} a_{1k_1} \cdots a_{pk_p} = \sum_K (-1)^{\rho_K} \lambda_1^{p-k_1} \lambda_2^{p-k_2} \cdots \lambda_p^{p-k_p}.$$

Here, ρ_K is the number of transpositions needed to bring $(k_1, \ldots, k_p)$ to the natural order $(1, 2, \ldots, p)$. Then, if ρ_K is odd then we have -1 and if ρ_K is even then we have $+1$ as the coefficient. For example, for $p = 3$ the possible permutations are $(1, 2, 3), (1, 3, 2), (2, 3, 1), (2, 1, 3), (3, 1, 2), (3, 2, 1)$. There are $3! = 6$ terms. In the general case there are $p!$ terms. For example, for the sequence $(1, 2, 3)$ we have $k_1 = 1, k_2 = 2, k_3 = 3 \Rightarrow \rho_K = 0$ and the corresponding sign is $+1$. For $(1, 3, 2)$ we have $k_1 = 1, k_2 = 3, k_3 = 2$. Here one transposition is needed to bring to the natural order $(1, 2, 3)$ and hence $\rho_K = 1$ and the corresponding sign is -1, and so on. In the complex case,

$$\prod_{i<j}(\lambda_i - \lambda_j)^2 = |A|^2 = |AA'| = |A'A|.$$

Let $A' = [\beta_1, \beta_2, \ldots, \beta_p]$, where β_j is the j-th column of A', $\beta_j' = [\lambda_j^{p-1}, \lambda_j^{p-2}, \ldots, \lambda_j, 1]$, $j = 1, \ldots, p$. Let $A'A = B = (b_{ij})$, $b_{ij} = \sum_{r=1}^p \lambda_r^{2p-(i+j)}$. Then,

$$\prod_{i<j}(\lambda_i - \lambda_j)^2 = |B| = |A'A| = \sum_K (-1)^{\rho_K} b_{1k_1} b_{2k_2} \cdots b_{pk_p}$$

where

$$b_{1k_1} = \lambda_1^{2p-(1+k_1)} + \lambda_2^{2p-(1+k_1)} + \cdots + \lambda_p^{2p-(1+k_1)}$$
$$b_{2k_2} = \lambda_1^{2p-(2+k_2)} + \lambda_2^{2p-(2+k_2)} + \cdots + \lambda_p^{2p-(2+k_2)}$$
$$\vdots = \vdots$$
$$b_{pk_p} = \lambda_1^{2p-(p+k_p)} + \lambda_2^{2p-(p+k_p)} + \cdots + \lambda_p^{2p-(p+k_p)}$$

Let
$$b_{1k_1}b_{2k_2}\cdots b_{pk_p} = \sum_{r_1,\ldots,r_p} \lambda_1^{r_1}\cdots\lambda_p^{r_p}. \tag{7.16}$$

In the real case, the joint density of $\lambda_1, \ldots, \lambda_p$ is the following:

$$g_3(D)\mathrm{d}D = \frac{\Gamma_p(\alpha + \frac{p}{2})}{\Gamma_p(\frac{n}{2})\Gamma_p(\alpha + \frac{p}{2} - \frac{n}{2})} \frac{\pi^{\frac{p^2}{2}}}{\Gamma_p(\frac{p}{2})} \left\{\prod_{j=1}^p \lambda_j^{\frac{n}{2} - \frac{p+1}{2}}\right\} \left\{\prod_{j=1}^p (1-\lambda_j)^{\alpha - \frac{n-1}{2}}\right\}$$
$$\times \left(\sum_K (-1)^{\rho_K} \lambda_1^{p-k_1} \lambda_2^{p-k_2} \cdots \lambda_p^{p-k_p}\right) \mathrm{d}D$$
$$= \frac{\Gamma_p(\alpha + \frac{p}{2})}{\Gamma_p(\frac{n}{2})\Gamma_p(\alpha + \frac{p}{2} - \frac{n}{2})} \frac{\pi^{\frac{p^2}{2}}}{\Gamma_p(\frac{p}{2})} \sum_K (-1)^{\rho_K} \lambda_1^{m_1}\cdots\lambda_p^{m_p} \left\{\prod_{j=1}^p (1-\lambda_j)^{\alpha - \frac{n-1}{2}}\right\} \mathrm{d}D \tag{7.17}$$

where, $\Re(\alpha) > \frac{n-1}{2}$ and $m_j = \frac{n}{2} - \frac{p+1}{2} + p - k_j$.

In the complex case,

$$\tilde{g}_3(D)\mathrm{d}D = \frac{\tilde{\Gamma}_p(\alpha + p)}{\tilde{\Gamma}_p(n)\tilde{\Gamma}_p(\alpha + p - n)} \frac{\pi^{p(p-1)}}{\tilde{\Gamma}_p(p)} \sum_K (-1)^{\rho_K} \lambda_1^{m_1}\cdots\lambda_p^{m_p} \left\{\prod_{j=1}^p (1-\lambda_j)^{\alpha - n}\right\} \mathrm{d}D \tag{7.18}$$

where $\Re(\alpha) > n - 1$ and $m_j = n - p + r_j$ where r_j is defined in (7.16). Hence, in (7.17) and (7.18) we will use the same notation m_j as the exponent of λ_j, $j = 1,\ldots,p$ with the understanding that in the real case $m_j = \frac{n}{2} - \frac{p+1}{2} + p - k_j$ and in the complex case $m_j = n - p + r_j$. Further, for simplicity, we may write the joint density of the eigenvalues in the real and complex cases as the following:

$$g_3(D)\mathrm{d}D = c \sum_K (-1)^{\rho_K} \lambda_1^{m_1}\cdots\lambda_p^{m_p}(1-\lambda_1)^\gamma\cdots(1-\lambda_p)^\gamma \mathrm{d}D \tag{7.19}$$

$$\tilde{g}_3(D)\mathrm{d}D = \tilde{c} \sum_K (-1)^{\rho_K} \lambda_1^{m_1}\cdots\lambda_p^{m_p}(1-\lambda_1)^\gamma\cdots(1-\lambda_p)^\gamma \mathrm{d}D \tag{7.20}$$

where

$$m_j = \frac{n}{2} - \frac{p+1}{2} + p - k_j, \gamma = \alpha - \frac{n-1}{2}, c = \frac{\Gamma_p(\alpha + \frac{p}{2})}{\Gamma_p(\frac{n}{2})\Gamma_p(\alpha + \frac{p}{2} - \frac{n}{2})} \frac{\pi^{\frac{p^2}{2}}}{\Gamma_p(\frac{p}{2})}, j = 1,\ldots,p$$

in the real case, and

$$m_j = n - p + r_j, \gamma = \alpha - n, \tilde{c} = \frac{\tilde{\Gamma}_p(\alpha + p)}{\tilde{\Gamma}_p(n)\tilde{\Gamma}_p(\alpha + p - n)} \frac{\pi^{p(p-1)}}{\tilde{\Gamma}_p(p)}, j = 1,\ldots,p$$

in the complex case, where r_j is defined in (7.16). Since we have written the joint density of the eigenvalues, both in the real and complex cases, by using the same

7.5 Exact Marginal Function of the Largest Eigenvalue λ_1 in (7.19)

format, we can use the same procedure to obtain the densities of the largest eigenvalue, smallest eigenvalue etc. Integration over $\lambda_1, \ldots, \lambda_{p-1}$ is needed to obtain the density of the smallest eigenvalue λ_p. Similarly, integration over $\lambda_p, \ldots, \lambda_2$ is needed to obtain the density of the largest eigenvalue λ_1. In the complex case, m_j, $j = 1, \ldots, p$ are always positive integers. Hence, integration by parts can get rid off the factor $\lambda_j^{m_j}$. But, if the m_j is large or moderately large then the final expression, even though a finite sum, will be messy. Similarly, when m_j or γ in the real or complex case is a positive integer, then integration by parts will eliminate the corresponding factor either $\lambda_j^{m_j}$ or $(1 - \lambda_j)^\gamma$. But the expressions may become messy when the parameter m_j or γ is large or moderately large. Hence, we will consider series expansions which will be valid for the real and complex cases. When the parameters are positive integers, then these series will terminate into a finite sum. Since $0 < \lambda_j < 1$ the series will converge fast even if it is an infinite series.

7.5 Exact Marginal Function of the Largest Eigenvalue λ_1 in (7.19)

For both the real and complex cases, whether γ is a positive integer or not, let us expand $(1 - \lambda_j)^\gamma$ to obtain a convenient representation. Note that

$$(1 - \lambda_p)^\gamma = \sum_{t_p=0}^{\infty} \frac{(-\gamma)_{t_p}}{t_p!} \lambda_p^{t_p}$$

where, for example, the notation $(a)_m = a(a+1)\cdots(a+m-1)$, $(a)_0 = 1$, $a \neq 0$ is the Pochhammer symbol. This will be a finite sum when γ is a positive integer. Now, we start integrating from λ_p onward.

$$\int_{\lambda_p=0}^{\lambda_{p-1}} \lambda_p^{m_p}(1-\lambda_p)^\gamma d\lambda_p = \sum_{t_p=0}^{\infty} \frac{(-\gamma)_{t_p}}{t_p!} \frac{1}{m_p + t_p + 1} \lambda_{p-1}^{m_p+t_p+1}.$$

Now, multiply this with $\lambda_{p-1}^{m_{p-1}}(1-\lambda_{p-1})^\gamma$ and integrate λ_{p-1} from 0 to λ_{p-2}, and so on. The final form, denoted by $f_1(\lambda_1)$ is the marginal function corresponding to the largest eigenvalue λ_1. That is,

$$f_1(\lambda_1) = \sum_{t_p=0}^{\infty} \frac{(-\gamma)_{t_p}}{t_p!} \frac{1}{m_p + t_p + 1} \sum_{t_{p-1}=0}^{\infty} \frac{(-\gamma)_{t_{p-1}}}{t_{p-1}!} \frac{1}{m_p + m_{p-1} + t_p + t_{p-1} + 2} \cdots$$

$$\times \sum_{t_2=0}^{\infty} \frac{(-\gamma)_{t_2}}{t_2!} \frac{1}{m_p + \cdots + m_2 + t_p + \cdots + t_2 + (p-1)} \lambda_1^{m_p+\cdots+m_1+t_p+\cdots+t_{p-1}+(p-1)}(1-t_1)^\gamma,$$

for $0 \leq \lambda_1 \leq 1$ and zero elsewhere. Now, incorporating the remaining factors from (7.17) and (7.18), we have the marginal density of λ_1.

7.6 Exact Marginal Function of λ_p, the Smallest Eigenvalue

In the complex case, m_j is a positive integer for all j, and in the real case, m_j is either a positive integer or a half-integer. Since we are integrating out, starting from $\lambda_1, \lambda_2 < \lambda_1 < 1$, we may write, for convenience,

$$\lambda_1^{m_1} = [1 - (1 - \lambda_1)]^{m_1} = \sum_{t_1=0}^{\infty} \frac{(-m_1)_{t_1}}{t_1!} (1 - \lambda_1)^{t_1}.$$

Then,

$$\int_{\lambda_1=\lambda_2}^{1} \lambda_1^{m_1} (1 - \lambda_1)^{\gamma} d\lambda_1 = \sum_{t_1=0}^{\infty} \frac{(-m_1)_{t_1}}{t_1!} \int_{\lambda_1=\lambda_2}^{1} (1 - \lambda_1)^{\gamma+t_1} d\lambda_1$$

$$= \sum_{t_1=0}^{\infty} \frac{(-m_1)_{t_1}}{t_1!} \frac{1}{\gamma + t_1 + 1} \lambda_2^{\gamma+t_1+1}.$$

Now, multiply by $\lambda_2^{m_2}(1 - \lambda_2)^{\gamma}$ and integrate out λ_2, and so on. Final result, denoted by $f_p(\lambda_p)$, is the following:

$$f_p(\lambda_p) = \sum_{t_1=0}^{\infty} \frac{(-m_1)_{t_1}}{t_1!} \frac{1}{\gamma + t_1 + 1} \sum_{t_2=0}^{\infty} \frac{(-m_2)_{t_2}}{t_2!} \frac{1}{2\gamma + t_1 + t_2 + 2} \cdots$$

$$\times \sum_{t_{p-1}=0}^{\infty} \frac{(-m_{p-1})_{t_{p-1}}}{t_{p-1}!} \frac{1}{(p-1)\gamma + t_1 + \cdots + t_{p-1} + (p-1)}$$

$$\times \lambda_p^{m_p}(1 - \lambda_p)^{p\gamma + t_1 + \cdots + t_{p-1} + (p-1)}, 0 \le \lambda_p \le 1$$

and this multiplied by the remaining factors from (7.17) and (7.18) will be the marginal density of the smallest eigenvalue λ_p. Observe that if one wishes to compute the marginal density of λ_s for any s, then integrate out $\lambda_1, \ldots, \lambda_{s-1}, \lambda_p, \ldots, \lambda_{s+1}$. Then, multiply by the remaining factors from (7.17) and (7.18). By integrating out $\lambda_1, \ldots, \lambda_{s-}$, one obtains the joint marginal function of $\lambda_s, \ldots, \lambda_p$, and so on.

References

Deppisch, F.F.: A Modern Introduction to Neutrino Physics, Morgan & Claypool Publishers. San Rafael, CA, USA (2019)

Fortin, J.-F., Giasson, N., Marleau, L., Pelletier-Dumont, J.: Mellin transform approach to rephasing invariants. Phys. Rev. D **102**, 036001 (2020)

Fortin, J.-F., Glasson, N., Marleau, L.: Probability density function for neutrino masses and mixings. Phys. Rev. D **94**, 115004-1–11500-13 (2016)

References

Haba, N., Shimizu, Y., Yamada, T.: Neutrino mass square ratio and neutrinoless double-beta decay in random neutrino mass matrices. Prog. Theor. Phys. 023B07 (2023)

Haba, N., Murayama, H.: Anarchy and hierarchy. Phys. Rev. D **63**, 053010 (2001). arXiv:hep-ph/0009174

Haubold, H.J., Mathai, A.M.: Does Super Kamiokande Observe Lévy Flights of Solar Neutrinos? (2024). https://doi.org/10.48550/arXiv.2405.11057

Mathai, A.M.: Jacobians of Matrix Transformations and Functions of Matrix Arguments. World Scientific, New York (1997)

Mathai, A.M., Haubold, H.J.: Erdélyi-Kober Fractional Calculus from a Statistical Perspective, Inspired by Solar Neutrino Physics, Springer Briefs in Mathematical Physics 31. Springer Nature, Singapore (2018)

Mathai, A.M., Provost, S.B., Haubold, H.J.: Multivariate Statistical Analysis in the Real and Complex Domains. Springer Nature, Switzerland (2022)

Oberauer, L., Ianni, A., Serenelli, A.: Solar Neutrino Physics: The Interplay between Particle Physics and Astronomy. Wiley-VCH, Weinheim, Germany (2020)

Scafetta, N.: Fractal and Diffusion Entropy Analysis of Time Series. Theory, concepts, applications and computer codes for studying fractal noises and Lévy walk signals. VDM Verlag Dr, Mueller, Saarbruecken (2010)

Yanagida, T.: Horizontal symmetry and mass of the top quark. Phys. Rev. D **20**, 2986 (1979)

Open Access This chapter is licensed under the terms of the Creative Commons Attribution 4.0 International License (http://creativecommons.org/licenses/by/4.0/), which permits use, sharing, adaptation, distribution and reproduction in any medium or format, as long as you give appropriate credit to the original author(s) and the source, provide a link to the Creative Commons license and indicate if changes were made.

The images or other third party material in this chapter are included in the chapter's Creative Commons license, unless indicated otherwise in a credit line to the material. If material is not included in the chapter's Creative Commons license and your intended use is not permitted by statutory regulation or exceeds the permitted use, you will need to obtain permission directly from the copyright holder.

Author Index

A
Abe, K., 129
Alpher, R.A, 85
Atkinson, R.D'E., 5, 19

B
Bahcall, J.N., 8, 19, 29, 51–53, 55, 56, 61, 63, 76
Barnes, C.A., 30, 38, 47, 64
Bemmerer, D., 127
Bertulani, C.A., 135
Bethe, H.A., 7, 8, 19, 81–83
Biswas, S., 19
Blatt, J.M., 21, 23, 24
Bohr, N., 4
Boltzmann, L., 58, 78
Breit, G., 7, 24, 25
Buldgen, G., 128, 135
Burbidge, E.M., 15–17, 19
Burbidge, G.R., 15–17, 19

C
Caughlan, G.R., 19–21, 23, 24, 26, 27
Chandrasekhar, S., 6–8, 52, 58, 60
Cheoun, M.-K., 135
Clausius, R., 1
Clayton, D.D., 62

Cleveland, B.T., 83
Condon, E.U., 5
Cravens, J.P., 129
Critchfield, C.L., 7, 19, 30, 37, 61, 63, 81
Culbreth, G., 130

D
Davis Jr., R., 8, 51, 55, 76, 80, 81, 84, 135
De Broglie, M.L., 24
Debye, P., 4

E
Eddington, A.S., 4
Emden, R., 3

F
Fowler, R.H., 3, 5, 19–21, 23, 24, 26, 27, 51, 52, 55, 63, 81, 82
Fowler, W.A., 7, 19–21, 23, 24, 26, 27, 51, 52, 55, 63, 81, 82

G
Gamow, G., 5, 19, 23, 25, 26, 64, 76
Gerth, E., 51, 71, 84
Gurney, R.W., 5

H
Hall, F.W., 135
Harmer, D.S., 8, 76
Harris, M.J., 8
Haubold, A., 19–22, 26, 28–35, 49, 51, 52, 56, 60–65, 69, 93, 97, 103, 124, 129, 131, 134, 135
Haubold, H.J., 19–22, 26, 28–35, 49, 51, 52, 56, 60–65, 69, 83, 87, 93, 103, 124, 129, 131, 134, 135
Haxton, W.C., 51, 53, 55
Hayashi, C., 7, 52, 60
Helmholtz. H., 71
Henyey, L.G., 7
Heo, K., 135
Hertzsprung, E., 4
Hilfer, R., 134
Hoffman, K.C., 8, 76
Houtermans, F.G., 5, 19
Hoyle, F., 7, 15, 17, 19
Hückel, E., 4
Huebner, W.F., 63
Hwang, E., 135

I
Iben Jr., I., 8

J
Jang, D., 135
John, R.W., 19, 26, 28–31, 33, 60, 63, 71, 76, 78, 80, 84

K
Kavanagh, R.W., 52, 53, 55, 62
Kippenhahn, R., 8
Klafter, J., 132
Ko, H., 135
Kramers, H.A., 5
Kumar, D., 135

L
Lane, J.H., 1, 2
Le Levier, R., 7
Lenee, R.D., 7
Lubow, S.H., 63

M
Mathai, A.M., 19–22, 26, 29–35, 38–40, 47–49, 52, 56, 60–65, 67, 69, 76, 78, 80, 83, 84, 87, 89, 93–95, 97, 99, 103, 107, 108, 110–112, 115–124, 129, 131, 134

Maxwell, J.C., 2, 26
Mayer, R.J., 1
Metzler, R., 132
Milne, E.A., 5

O
Öpik, E.J., 7
Orebi Gann, G.D., 127

P
Parker, P.D., 60, 61, 63
Perrin, J., 4
Piatibratova, O., 135
Pontecorvo, B., 8, 75
Provost, S.B., 124

R
Ramadurai, S., 19
Ritter, A., 1, 2
Rowley, J.K., 55
Russell, H.N., 4

S
Sakurai, K., 78, 127
Salpeter, E.E., 7, 19, 24, 61, 63, 81
Sampson, R.A., 3
Saxena, R.K., 29, 34, 38–40, 47, 48, 52, 63, 65, 67, 76, 80
Scafetta, N., 84, 128, 130
Schönberg, M., 7
Scholkmann, F., 128
Schuster, A., 3
Schwarzschild, K., 3
Schwarzschild, M., 7
Serenelli, A., 127
Slad, L.M., 135
Strömgren, R., 6
Slad, L.M., 135
Sturrock, P.A., 128, 135
Sugimoto, D., 7, 52, 60

T
Teller, E., 19
Thomson, W. (Lord Kelvin), 1
Tian, Z., 135
Tsallis, C., 116, 131

Author Index

U
Ulrich, R.K., 63

V
Vahla, M.N.S., 19
Vogt, H., 5

W
Wagoner, R.V., 26, 27
Weisskopf, V.F., 21, 23, 24
Weizsäcker, C.F. v., 7, 8

Wentzel, G., 5
Wigner, E., 7, 25
Wilson, A.H., 5

Y
Yang, W., 135
Yoo, J., 128

Z
Zimmerman, B.A., 19–21, 23, 24, 26, 27
Zuber, K., 127

Subject Index

C
Coulomb barrier, 5, 15, 18, 25, 56, 63
Cross section factor, 7, 9, 24, 27, 29, 33, 55, 56, 61, 63

D
Diffusion entropy analysis, 84, 86, 89, 128, 130, 131, 135, 137

E
Equation of state, 5, 6, 9, 52, 60
Erdélyi-Kober fractional integral, 120–124
Euler transform, 117

F
Fractional integral, Mellin transform, generalized, 119
Fractional integral, scalar case, first kind, 123

G
Gamow penetration factor, 5

H
Hep neutrinos, 84
Homestake experiment, 75, 76, 78, 84

K
Kamiokande experiment, 135
Krätzel transform, 83, 89, 97

M
Mathai entropy, 116
Meijer function, 29–31, 37, 43, 47, 62, 65, 69
MSW effect, 88, 89

N
Neutrino, astrophysics, 71, 75, 79, 89, 137
Neutrino, oscillations, 2, 75, 84, 88, 135, 137

Neutrinos, solar, 7–10, 51, 52, 55, 59, 64, 69, 71, 75–80, 83–89, 127–133, 135, 137
Neutrinos, variation with time, 127
Nuclear, astrophysics, 15, 19, 89, 90, 93, 94, 96, 98, 100, 102, 104, 106, 108, 110, 112, 114, 116, 118, 120, 122, 124
Nuclear, reactions, 5–8, 15, 19, 21–27, 35, 46, 51, 55, 56, 60–62, 70, 82, 83, 90, 93, 97

P
Pathway model, 99
Pep neutrinos, 87, 127
Pochhammer symbol, basic, second kind, complex case, 149
PP chain, 7, 8, 87, 127

PP neutrinos, 87
Probability density function, 86, 87, 89, 90, 128, 130, 132, 135, 137
Product, real scalar case, 120

R

Ratio, real scalar case, 120
Reaction rates, 6, 8–10, 19, 20, 22, 25–29, 32, 35, 37, 52, 55, 60–63, 70, 81
Riemann-Liouville fractional integral, left-aided, 120

S

Solar neutrino problem, 51
Solar neutrinos, 8, 10, 51, 52, 55, 59, 64, 69, 71, 75–80, 83–89, 127–137
Standard deviation analysis, 86, 89, 127, 128, 132, 133, 135
Stellar energy generation, 7, 19, 20, 51, 52, 54
Sum, real scalar case, 118
Sun, 1, 3, 4, 7–9, 51, 52, 55–60, 62, 64, 69, 70, 76, 84, 88–90, 127

W

Weyl fractional integral, second kind, 120

MIX
Papier aus verantwortungsvollen Quellen
Paper from responsible sources
FSC® C105338

If you have any concerns about our products,
you can contact us on
ProductSafety@springernature.com

In case Publisher is established outside the EU,
the EU authorized representative is:
**Springer Nature Customer Service Center GmbH
Europaplatz 3, 69115 Heidelberg, Germany**

Printed by Libri Plureos GmbH
in Hamburg, Germany